A Practical Guide
to
Textile Testing

A Practical Guide to Textile Testing

K. Amutha

Published by Woodhead Publishing India Pvt. Ltd.
Woodhead Publishing India Pvt. Ltd.,
303, Vardaan House, 7/28, Ansari Road,
Daryaganj, New Delhi - 110002, India
www.woodheadpublishingindia.com

First published 2016, Woodhead Publishing India Pvt. Ltd.
© Woodhead Publishing India Pvt. Ltd., 2016

Woodhead Publishing India Pvt. Ltd. ISBN: 978-93-85059-07-0
Woodhead Publishing India Pvt. Ltd. e-ISBN: 978-93-85059-06-9

Typeset by Mind Box Solutions, New Delhi
Printed and bound by Replika Press Pvt. Ltd.

Contents

Preface

Textile can be a fascinating term to mankind because of aspects such as colour, texture, design and comfort involved in its usage. The use of textiles by humans began with the identification of fibre, which dates back to prehistoric times. Such textiles are available in different forms for various end-uses like apparel, home textiles, and technical textiles. Here comes the necessity for testing of these textiles so as to ensure the quality of the product. Testing can be carried out at different stages, beginning from the raw material – fibre, and the subsequent intermediaries such as yarn, fabric – grey and processed stages, and finally, the garment.

Testing needs to be carried out in a well-organized manner since test results are used for evaluating product or fabric quality. Hence, given the importance of testing, various testing methods and procedures are standardized by organizations such as ISO, AATCC, ASTM, BSI, DIN, ANSI, and so on. The testing standards set by these institutions are unique and developed after careful research. It is crucial to understand the importance and necessity of textile testing. It is necessary that aspiring professionals and readers of this book understand the implications of terminologies such as calibration, reliability, repeatability and traceability, as they represent key criteria, parameters, and deliverables expected to be achieved via testing.

The aim of this book is to give specific information about the various procedures involved in textile testing in order for learners to gain knowledge about practical approaches utilized in textile testing. The standard atmosphere for testing, influence of moisture on properties of textiles, sampling methods' importance as well as conditioning of sample before testing, testing procedures and, finally, the evaluation of results are explained.

This book is divided into six parts: First, introduction to textile testing with the sources of testing standards, sampling for testing, moisture and its relation with textiles; second, fibre testing; third, yarn testing; fourth, fabric testing; fifth, testing for export market; and sixth, accreditation of textile testing laboratory. Each chapter is self-explanatory, and on the whole, the book is a complete guide to textile testing.

I feel honoured to author this book, which is a collection of my experiences in textile testing, and to publish this with Woodhead Publishing India, a leading and eminent publisher in textile technology. My sincere thanks to Ms. Harpreet Kaur for her consistent efforts towards this publication. Above all, I thank Lord Almighty, my family and colleagues. I hope the book is informative and useful to the readers.

Amutha, K.

1

Introduction

Definition: Applying engineering knowledge and science to detect the criteria and properties of any textile material or product (such as fibre, yarn, fabric) is called *textile testing*.

Objectives of testing

- To check the quality and suitability of raw material
- To monitor the production (process control)
- To assess the quality of final product
- To investigate the faulty materials
- To set standards or benchmarks
- For R&D (research and development) purpose
- For new product development

Importance of Testing

- To ensure the product quality
- To control the manufacturing process
- For customer satisfaction and retention
- Good reputation (brand image) among consumers

1.1 Testing methods (sources of testing standards)

Testing is done primarily to test the quality and there are different ways to carry out a test. Sometimes, different principles and instruments may be employed to test a single criterion. Hence it is important to standardize the testing methods or procedures. Various national and international organizations have established standards for textile testing. Some of the organizations involved in developing textile testing standards are as follows:

- AATCC - American Association of Textile Chemists and Colorists
- ASTM - American Society for Testing and Materials
- ANSI - American National Standards Institute
- ISO - International Organization for Standardization
- BSI - British Standards Institute

- BIS - Bureau of Indian Standards
- BS EN - British Standard European Norm
- IS - Indian Standards

1.2   Selection of samples for testing

Sample: It is a relatively small fraction selected from a population; the sample is supposed to be a true representative of the population.

Population: All elements, individuals or units that meet the selection criteria for a group to be studied and from which a representative sample is taken for detailed examination. It is the total system that need to be studied.

Need for sampling: Textile testing is destructive in nature, i.e. the materials used for testing go as waste after testing and hence it is not desirable to test all of the material. As textile production is always huge and bulk it is impossible to test all the final output from a production process. Thus, only representative samples of the material are tested. Sampling saves time and cost.

Sampling methods depends on the following factors:

- Form of the material
- Amount of material available
- Nature of the test
- Type of testing instrument
- Information required
- Degree of accuracy required

Types of sample

Random sample: Every individual in the population has an equal chance of being selected as a sample. It is free from bias, therefore it is a true representative of the population.

Numerical sample: A sample in which the proportion by number of, say, long, medium and short fibres, would be the same in the sample as in the population.

Biased sample: When the selection of an individual is influenced by factors other than chance, a sample ceases to be truly representative of the bulk and leads to bias in results.

Causes of bias in sampling

1. *Bias due to physical characteristics:* Longer fibres have a greater chance of being selected.

Position relative to the person: Lab assistant may pick bobbins from the top layer of a case of yarn (just to make his job easier or may be because of his ignorance), but the bobbin chosen will be biased due to their position.

2. ***Subconscious bias:*** Person selecting cones will pick the best-looking ones that are free from ridges, cub webbed ends and so on. This affects the test results.

1.3   Sampling for textile testing

- Fibre stage
- Yarn stage
- Fabric stage
- Garment stage

1.4   Fibre sampling techniques

Sampling of raw cotton

Since 100% testing of fibre is not possible, random sampling is done.

Zoning technique: A sampling method for cotton fibres

As cotton in bulk is not homogeneous, a number of sub-samples must be taken at random from different places in the bulk. When samples are drawn from cotton bales, the required amount of fibres should be taken one by one at random from different parts of the bale.

- Step 1: A sample that weighs 2 ozs (approximately 906.72 gm) is drawn by selecting about 80 large tufts from different parts of the bulk.
- Step 2: This sample is then divided into four parts.
- Step 3: Sixteen small tufts are taken at random from each part (approximately 20 mg).
- Step 4: Each tuft is halved four times, discarded alternately by turning the tuft through right angle between successive halving. Sixteen wisps are thus produced from each part.
- Step 5: These wisps are combined to form a tuft.
- Step 6: Each tuft is mixed by doubling and drawing between fingers.
- Step 7: Each tuft is divided into four parts.
- Step 8: A new tuft is obtained by combining a part of each of four tufts.

- Step 9: Sample is mixed again by doubling and drawing.
- Step 10: A quarter of sample is taken out from each tuft to form final sample.

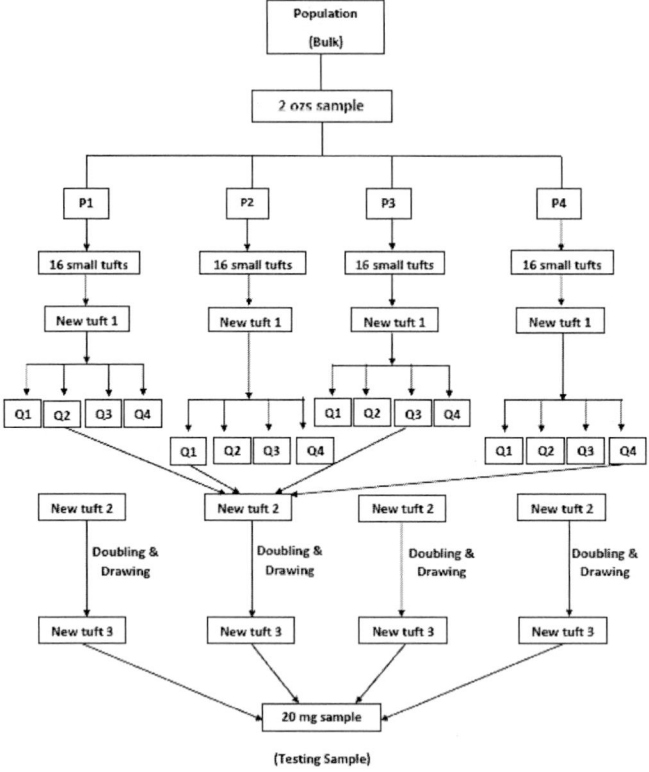

Figure 1.1 Zoning technique

Core sampling: A sampling method for wool fibres

Core sampling is a common method for obtaining a laboratory sample of clean wool from a lot of packaged raw wool. This method of sampling is done to assess the proportion of foreign matters such as grease and vegetable matter in unopened bales of raw wool.

Procedure

Weigh the bale just before coring process. Make a small hole in the bale cover and plunge the coring tube either manually or by drilling. The tube is entered in the direction of compression so that the cut is perpendicular to the layers of the fleece.

The bale may be divided into eight segments of approximately equal volume. Collect fibre samples from different segments of the bale, in different directions so that a wide array of fibres are collected. The depth of penetration has to be maintained at same level for each core in a given lot.

About 2.5 pounds (ozs) of fibre sample has to be collected by core sampling process. The number of cores obtained depends on the dimensions of the coring tube.

The coring tube is narrow with dimensions of 2 feet length and 0.75 inches diameter. It has a sharp cutting tip at one end and a pair of handles at the other end. As the coring tube enters the bale a plug of material is forced inside the tube. In order to collect the core collected in the tube a slit and blade arrangement is being provided by the side of the tube. The core so ejected from the coring tube is collected in a bag provided at the top end of the coring tube. A number of such cores are collected and used as a representative sample for testing.

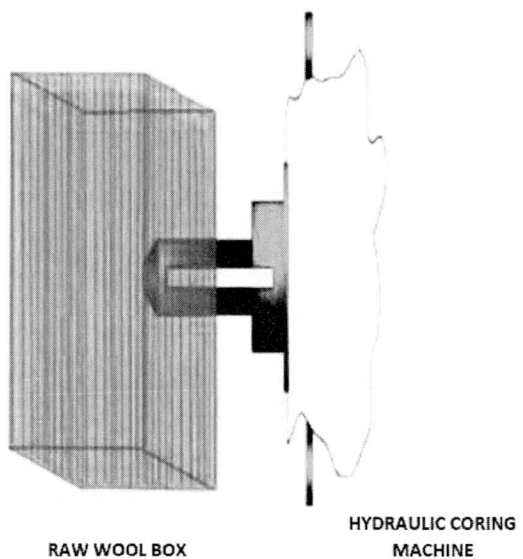

RAW WOOL BOX   HYDRAULIC CORING MACHINE

Figure 1.2 Core sampling

Fibre sampling from combed slivers or roving or yarns

- Random draw method
- Cut square method
- Random draw method

This method is used for sampling card sliver, ball sliver and top. The sliver from which sample has to be taken is pulled in such a way that the end has no broken or cut fibres. Then the sliver is kept on two velvet boards with the pulled end at the front of the first board. A glass plate is kept over the other end of the sliver so that it remains in its position and does not move.

Then using a wide grip, 2 mm fringes of fibre are removed from the front end of the sliver and discarded. This process of removing and discarding fibres is repeated until a distance equal to the longest fibre in the sliver has been removed.

The front end of the sliver is now 'normalised', and further drawing of fibre results in a sample of fibres representing different lengths of fibre available in the sliver. As these fibres tend to be a numerical sample all the fibres that lie between two lines are taken as the sample.

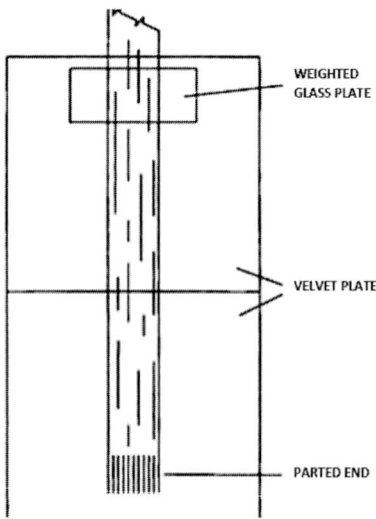

WEIGHTED
GLASS PLATE

VELVET PLATE

PARTED END

Figure 1.3 Random draw method

Cut square method

This method is used for obtaining fibre sample from yarn. Cut a certain length of the yarn and then untwist one of the ends of the yarn by hand. Then lay the untwisted yarn on a small velvet board and cover with a glass plate. Then cut the untwisted end of the yarn at about 5 mm from the edge of the plate. Remove all the fibres that project in front of the glass plate one by one with a pair of forceps and discard.

Now, all the cut fibres would be removed, leaving only the uncut fibres with their original length. Then move the glass plate back a few millimetres, exposing more fibre ends. Again remove these fibres one by one and measure. When all the fibre lengths have been measured move the plate back again until a total of 50 fibres have been measured. In each case, once the plate is moved all projecting fibre ends must also be removed and measured. The whole process is then repeated on fresh lengths of yarn chosen at random from the bulk, until sufficient fibres have been measured.

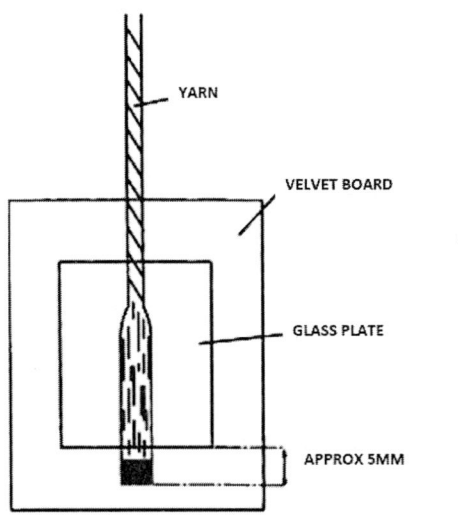

Figure 1.4 Cut square method

1.5   Yarn Sampling Techniques

Random sampling - yarn in package form

Yarn is available in various forms of package such as bobbins, cops, cone and cheese and as hanks. Table of random number is normally used sampling yarn bobbins from comparatively small bulk size. Totally 10 packages may be selected at random.

(a)   If the bulk contains more than five cases, at least five cases are selected at random and then two packages are selected at random from each case.

(b)   If the number of cases is less than five, then ten packages are selected at random approximately, two from each package.

1.6   Fabric sampling techniques

Figure 1.5 shows correct sampling method for woven fabric. Fabric samples from warp and weft are taken separately as their properties vary substantially along warp and weft. Identify and mark the warp direction first. Make sure that no two specimens contain same warp or weft threads. Mark and cut samples at least 2 inches away from the selvedge. Also, make sure not to take samples from creased, wrinkled or damaged portions of the fabric, if any. In case of knit fabric, samples are taken from different parts of the fabric almost the same same way as done for wovens.

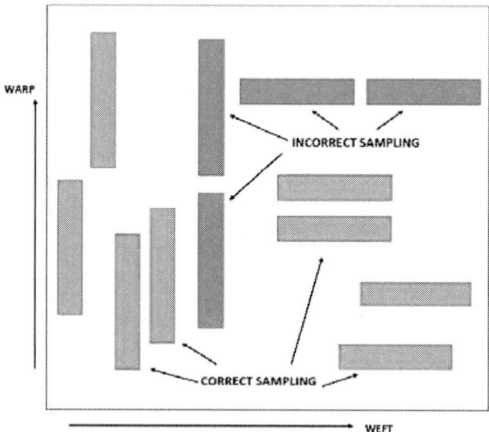

Figure 1.5 Fabric sampling

1.7   Standard atmosphere for testing

Moisture equilibrium. It is the condition reached by a material when it no longer takes up moisture from, or gives up moisture to, the surrounding atmosphere.

Pre-conditioning. To bring a sample or specimen of a textile to relatively low moisture content (approximate equilibrium atmosphere with relative humidity between 5% and 25%) prior to conditioning in a controlled atmosphere for testing.

Conditioning. To bring a material to moisture equilibrium with a specified atmosphere. Before a textile is tested, it is conditioned by placing it in the atmosphere for testing in such a way that the air flows freely through the textile and keeping it there for the time required to bring it into equilibrium with the atmosphere. Unless otherwise specified, the textile should be considered to be in equilibrium when successive weighing, at specific time intervals, shows no progressive change in mass greater than 0.25%.

Standard atmosphere for testing textiles. Laboratory conditions for testing fibres, yarns and fabrics in which air temperature and relative humidity are maintained at specific levels with established tolerances. Textile materials are used in a number of specific end-use applications that frequently require different testing temperatures and relative humidity. Specific conditioning and testing of textiles for end-product requirements can be carried out using table 2.6.

Table 2.6 Standard atmospheres for testing various materials

Material	Temperature	Relative Humidity %
Textiles other than nonwoven, Tire cords and glass fibre	$21 \pm 1^\circ$ C	65 ± 2
Nonwovens (includes paper)	$23 \pm 1^\circ$ C	50 ± 2
Glass fibre for textile applications	$21 \pm 1^\circ$ C	65 ± 5

Atmospheric conditions and relative humidity. The dampness of atmosphere can be calculated in terms of humidity.

Absolute humidity. The weight of water present in a unit volume of moist air, that is, gm/m^3.

Relative humidity. The ratio of the absolute humidity of the air to that of air saturated with water vapour at the same temperature and pressure, expressed as a percentage.

RH% = (Absolute humidity of air / Humidity air saturated with water vapour) × 100

Measurement of R.H. percentage: A hygrometer or psychrometer is an instrument used for measuring the moisture content in the atmosphere. A psychrometer, or wet-and-dry-bulb thermometer, consists of two thermometers, one that is dry and one that is kept moist with distilled water on a sock or wick. The two thermometers are thus called the dry bulb and the wet bulb. At temperatures above the freezing point of water, evaporation of water from the wick lowers the temperature, so that the wet-bulb thermometer usually shows a lower temperature than that of the dry-bulb thermometer. When the air temperature is below freezing, however, the wet bulb is covered with a thin coating of ice and may be warmer than the dry bulb.

Relative humidity is computed from the ambient temperature as shown by the dry-bulb thermometer and the difference in temperatures as shown by the wet-bulb and dry-bulb thermometers. Relative humidity can also be determined by locating the intersection of the wet- and dry-bulb temperatures on a psychrometric chart. The two thermometers coincide when the air is fully saturated, and the greater the difference the drier the air.

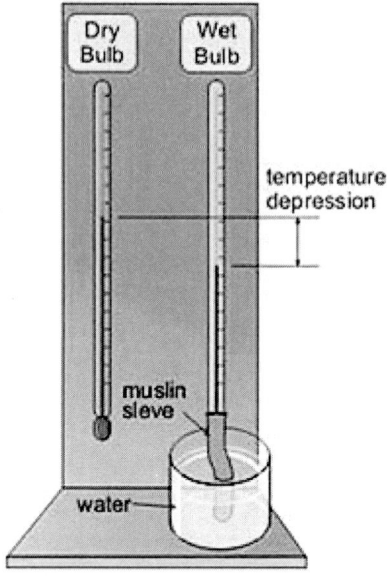

Figure 1.6: Hygrometer

For example,

Dry-bulb reading = 68°F

Wet-bulb reading = 61°F Difference = 7°F

R.H. percentage from chart = 67%

With advancement in technology, digital hygrometers have also been developed and made available. They are simple to use and produce quick results.

Importance of moisture measurement

Moisture content of cotton makes significant changes in the physical properties of cotton and hence moisture content has to be known. High moisture content increases flexibility, toughness, elongation and tensile strength. If the moisture content is too high it causes difficulty in processing due to the tendency of the stock to lap-up on drafting rolls. Low moisture, on the other hand, facilitates cleaning but increases the brittleness of the fibre and results in fibre breakage during ginning, cleaning and mill processing. Low moisture also increases fly waste and may cause manufacturing difficulties due to static electricity.

1.8   Measurement of moisture regain

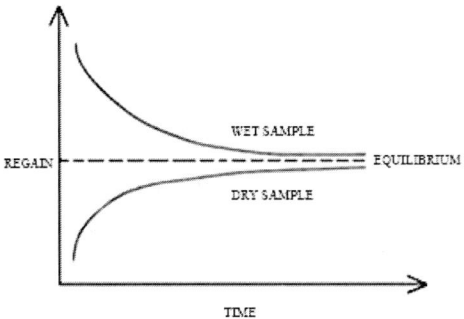

Figure 1.7 Moisture equilibrium

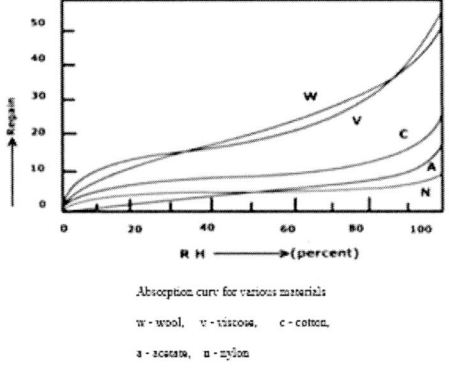

Absorption curv for various materials

w - wool,   v - viscose,   c - cotton,

a - acetate,   n - nylon

Figure 1.8 Absorption curves of textile fibres

Conditioning oven

This instrument is used for the determination of the amount of moisture in cotton by oven-drying and is applicable to raw cotton, cotton stock in process and cotton waste. This may also be used for determining moisture in blends of cotton with other fibres.

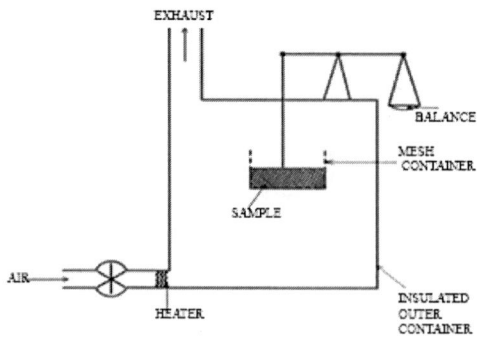

Figure 1.9 Conditioning oven

A conditioning oven is shown in Figure 1.9. It has a mesh container in which the fibre sample is placed. The mesh container acts as one on the pans of a weighing balance and the other pan is outside the oven. This set up ensures the weighing of the sample without any disturbances in the system. The fibre sample is placed in the mesh container and weighed. Then dry air is passed through the oven at a constant rate. Temperature of the air is maintained at $105 \pm 3°C$. Then the sample is weighed successively after time intervals of, say 20 minutes. When successive weighings differ less than 0.05% it may be assumed that a constant weight has been reached. The weight of moisture is the difference between the weight before drying (original weight) and the oven dry weight.

The main advantage of the conditioning oven is that all the weighing is carried out inside the oven and hence the accuracy is ensured. The method is based on the assumption that the air drawn into the oven is at the standard atmospheric condition. If not, then correction has to be made.

Moisture regain, MR = (W / D) × 100%

where W = weight of moisture; D = Oven dry weight of sample

Moisture content, MC = [W / (W+D)] × 100%

$$= MR / [1 + (MR/100)]$$

where W = weight of moisture; D = oven dry weight of sample, W + D = original weight of sample.

Shirley moisture meter

The electrical properties of fibres change markedly with the difference in moisture content and hence the measurement of resistance or capacitance changes can be used as an indirect method to measure regain.

The Shirley moisture meter has two electrodes with a non-conducting material in between the electrodes. The electrodes are in various sizes that enables to test materials in different forms such as bales of cotton, yarn packages, etc. The electrodes are plunged into a package of yarn and the resistance between the electrodes is measured. This electrical resistance is converted as moisture regain values and is displayed.

Since the instrument is used for different fibres and forms it has to be calibrated for each type of fibre. The great advantages of the electrical methods over drying and weighing methods are the speed, ease of reading and portability. The disadvantages of electrical methods are the need to recalibrate them as they are indirect methods, variations in readings due to packaging density, presence of dyes, anti-static agents and also variations in fibre quality.

Moisture and fibre properties

Dimensions. Absorption of moisture causes swelling of fibre and as a result shrinkage occurs in fabric. This could be taken advantageous in the design of waterproof fabrics.

Mechanical properties. Generally, moisture absorption weakens the fibre, but vegetable fibres such as cotton and flax are exceptional and their strength increase with absorption of moisture. Other mechanical properties like extensibility, crease recovery, flexibility and ability to be 'set' by finishing processes are affected by regain values.

Electrical properties. The high ratio of electrical resistance of textile fibre at low and high regain helps in the design of moisture meters. Dielectric and static characteristics are also affected by the amount of moisture in the material.

Thermal effect. Absorption of moisture by the fibre results in generation of heat which is referred to as 'heat of absorption'. This property of the textile fibre helps the wearer to withstand the sudden change in temperature and relative humidity, especially during winter.

Factors affecting the regain of textile material

Time. When a textile material is placed in a given atmosphere it takes a certain amount of time to reach equilibrium. This rate of conditioning depends on factors such as the size and form of material and the nature of the material, external conditions, etc.

Relative humidity. The regain of textile material depends on the relative humidity of the atmosphere. Regain is higher at higher relative humidity

(RH). This is well understood by the absorption-desorption curves as shown in Figure 1.8.

Temperature. It has negligible effect on regain. For example, a change of 10°C may bring a change of 0.3 percent in regain of cotton.

The previous history of sample. Regain is effected by the nature of the material and the atmospheric condition in which the material has been stored or processed. For example, bleached or scoured cotton will absorb more moisture than untreated material because the removal of impurities helps in more absorption of moisture.

2
Fibre testing

Cotton fibre length. Length of staple fibre is one of the most important characteristics. In general a longer average fibre length is to be preferred because it confers a number of advantages.

- First, longer fibres are easier to process.

- Second, more even yarns can be produced from them because there is less number of fibre ends in a given length of yarn.

- Third, a higher-strength yarn can be produced from them for the same level of twist.

2.1   Fibre length measurement

Baer sorter/comb sorter method

Comb sorter is used to determine the length of the fibre. Length is the most important property of a fibre. Comb sorter can be used with cotton, wool, viscose or polyester yarn/fibre to determine its length. Cumulative fibre length distribution is determined. Effective length, mean length, percentage of short fibres and percentage of dispersion are other important parameters determined by this method.

Merits

1. Effective length is close to grader's staple length.

2. Provides accurate estimate of short fibre content.

Limitations

1. Time consuming (2 hours per sample).

2. Calls for considerable operator skill in sampling and preparing the diagram.

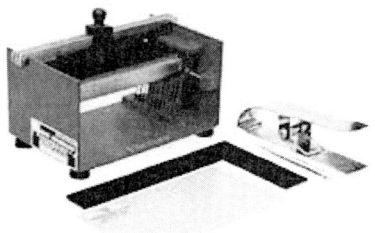

Figure 2.1 Comb sorter apparatus

Sample preparation

A representative sample of cotton is made into a sliver by drawing and doubling several times with the fibres straightened and parallelized. The bundle of fibres must be as narrow as possible throughout the whole process.

Procedure

- The sorter is placed with the back facing to the operator. The prepared sample is slightly pressed and placed on the bottom combs at the right-hand side of the sorter with a small portion half protruding.

- From the protruding end all the loose fibres are removed by means of tweezers, until ends are aligned. The removed loose fibres are kept separately and introduced in the original sample later.

- A tuft of fibres are pulled out, combed and transferred to the left-hand side of the sorter, so that the comb is nearest to the operator from the starting line for the tuft while at the other end the longer fibres protrude out. This tuft is pressed into the combs by means of depression.

- The process is repeated till all the fibres on the right-hand side are transferred to the left side. The top combs are inserted in their position to grip and control the slippage of fibres.

- The sorter is then turned around so that the front faces the operator.

- The bottom combs are dropped one by one successively till the tips of the longest fibres are seen.

- The fibres are pulled by the tweezers, combed, straightened and laid perpendicular to the baseline on the black velvet pad. When these fibres are exhausted, one more comb is dropped and fibres are fixed in the order of similar lengths, pulled once and laid on the pad. All the fibres are carefully spread out with uniform density and the process is continued until the tuft is exhausted and the entire fibre array is obtained.

- Later a pattern is built using a transparent scale rectangle-shaped with one side marked with 1/8" lines (Y axis) and the other side marked with ½" lines (X axis).

- Using the readings on the transparent scale, the values of the co-ordinates are marked on the graph sheet and the pattern is drawn. This diagram is called 'sorter diagram'. This diagram is analyzed for the following.
  - ○ Effective length
  - ○ Mean length
  - ○ Percentage of short fibres
  - ○ Dispersion of fibre length.

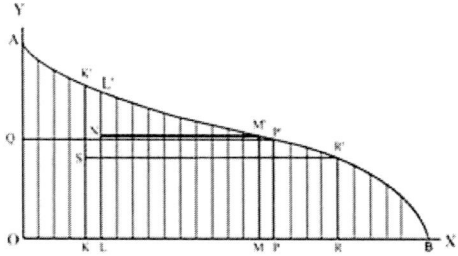

Figure 2.2 Comb sorter graph

Analysis of the sorter diagram;

Q is the midpoint of OA, that is, OQ = 1/2 OA.

From Q, QP' is drawn parallel to OB to cut the curve at P'.

PP' is drawn perpendicular to OB.

K is marked on OB, such that OK = ¼ OP and the perpendicular line KK' is drawn.

S is the midpoint of KK'.

From S, SR› is drawn parallel to OB to cut at B'.

The perpendicular line RR› is drawn to OB.

L is marked on OB, such that OL = % OR.

From L a perpendicular line LL' is drawn to cut the curve at L' (LL' = effective length).

Mark M, such that OM = $^3/_4$ OR and draw MM' perpendicular to OB to cut curve of M'.

Draw NM' parallel to OP to cut at N.

In the diagram

OQ = 1/2 OA

OK = 1/4 OP

KS = 1/2 KK'

OL = 1/4 OR.

Short fibre percentage = (RB/OB) × 100%.

LL' = Effective length (because many machine settings are related to this length).

Effective length is a characteristic of the bulk of the longer fibres.

Mean length: This is the average length of fibres in the sample. It is calculated as follows:

Mean length = (Area under the curve OAB) / OB

Table 2.1 Classification of fibre based on mean fibre length

Category	length (mm)
Short-B	Less than 19
Short-A	20.0 to 21.5
Medium	22.0 to 24.0
Long	24.5 to 26.0
Extra long	More than 27.0

Source: CICR: Central Institute for Cotton Research

Span length: Span length is measured with the help of digital fibrograph; 2.5% span length and 50% span length are determined. Span length is considered as standard in international market.

Table 2.2 Classification of fibre based on 2.5% span length

Category	2.5% Span length (mm)
Short-B	Less than 20
Short-A	20.5 to 24.5
Medium	25.0 to 29.0
Long	29.5 to 32.5
Extra long	More than 33.0

Source: CICR: Central Institute for Cotton Research

2.2   Fibre fineness measurement

A cotton fibre is a single elongated cell that grows from the epidermis of the cotton seed. Cotton fibre fineness is defined in terms of linear density, for example, in milligrams/kilometre (millitex - mtex)

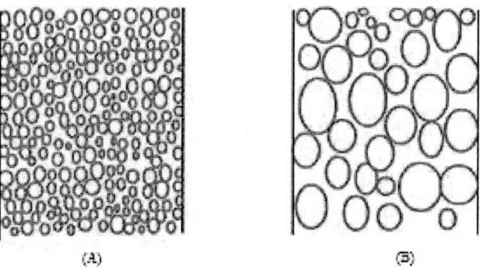

(A)   (B)

AIR FLOW THROUGH COARSE AND FINE FIBERS

Figure 2.3 Fine and coarse fibres

Fibre fineness influences primarily the following:

- Spinning limit
- Yarn strength
- Yarn evenness
- Yarn fullness
- Drape of the fabric
- Lustre
- Handle
- Productivity of the process

Importance of fibre fineness

It has been known since long that fibre fineness plays an important role in determining the quality of resultant yarn and hence that of the resultant fabrics. In general fibre fineness is important due to the following factors:

1. *It affects stiffness of the fabric*

- As the fibre fineness increases, resistance to bending decreases.
- It means the fabric made from yarn of finer fibre is less stiff in feel.
- It also drapes better.

2. *It affects torsional rigidity of the yarn*

- Torsional rigidity means ability to twist.
- As fibre fineness increases, torsional rigidity of the yarn reduces proportionally. Thus, fibres can be twisted easily during spinning operation.
- Also there will be less snarling and kink formation in the yarn when fine fibres are used.

3. *Reflection of light*

- Finer fibres also determine the lustre of the fabric.
- Because there are so many number of fibres per unit area they produce a soft sheen.
- This is different from hard glitter produced by coarser fibres.
- Also, the apparent depth of the shade will be lighter in case of fabrics made with finer fibres than those made with coarser fibres.

4. *Absorption of dyes*

- The amount of dye absorbed depends on the amount of surface area accessible for dye, out of a given volume of fibres. Thus, a finer fibre leads to quicker exhaustion of dyes than coarser fibres.

5. *Ease in spinning process*

- A finer fibre leads to more fibre cohesion because the number of contact surfaces are more and hence cohesion due to friction is higher.

- Also finer fibres lead to less amount of twist because of the same increased force of friction.

- This means yarns can be spun finer with the same amount of twist as compared to coarser fibres.

6. *Uniformity of yarn and hence uniformity in the fabric*

- Uniformity of yarn is directly proportional to the number of fibres in the yarn cross-section.

- Hence, finer the fibre, more uniform is the yarn. When the yarn is uniform it leads to other desirable properties such as better tensile strength, extensibility and lustre.

- It also leads to fewer breakages in spinning and weaving.

Cotton fineness measurement by air-flow principle (Sheffield Micronaire):

The resistance offered to the flow of air through a plug of fibres is dependent on the specific surface area of the fibres. Fineness tester has been developed on the basis of this principle for determining fineness of cotton.

Procedure

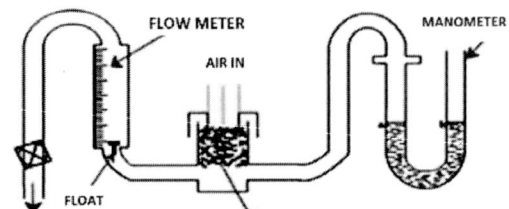

Figure 2.4 Sheffield micronaire tester

In the micronaire instrument, a weighed quantity of 3.24 gm of well-opened cotton sample is compressed into a cylindrical container of fixed dimensions. Compressed air is forced through the sample at a definite pressure and the volume rate of flow of air is measured by a rotometer-type flow meter. The sample for micronaire test should be well-opened, cleaned and thoroughly mixed (by hand-fluffing and opening method). Out of the various air-flow instruments, the micronaire is robust in construction, easy to operate and presents little difficulty as regards its maintenance.

Air flow α 1/S

Specific surface area (S) = π **d l** / π **d2** / **4** × **1** α **1** /**d**

By measuring the rate of air flow under controlled conditions, the specific surface area(s) of fibre can be determined and, consequently, the fibre diameter (also the fibre weight/unit length). The micronaire tester can be set at two different conditions:

  a. Measurement of air flow at a constant pressure drop.

  b. Measurement of pressure drop at a constant air flow.

Fibre fineness: In the international market unit of fibre fineness is millitex. Millitex is 37.39 times higher than micronaire. Fibre fineness is classified based on micronaire as follows and unit of fibre fineness is (microgram/inch).

Table 2.3 Classification of fibre based on fineness

Category	Fibre fineness (microgram/inch)
Very fine	Less than 3.0
Fine	3.0 to 3.9
Average	4.0 to 4.9
Coarse	5.0 to 5.9
Very coarse	More than 6.0

Source: CICR: Central Institute for Cotton Research

2.3   Cotton maturity measurement

The cotton fibre consists of cell wall and lumen. The maturity index is dependent upon the thickness of this cell wall. Cotton fibre maturity and the degree of secondary cell wall thickening relative to the perimeter are one of the most important fibre qualities and processing parameters of cotton. Immature fibres have neither adequate strength nor adequate longitudinal stiffness; therefore, they lead to loss of yarn strength, neppiness, a high proportion of short fibres, varying dyeability, processing difficulties, mainly at the card.

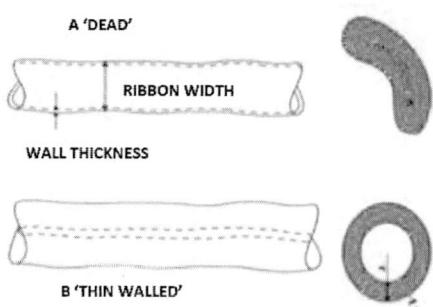

Figure 2.5 Cotton fibre maturity

Cotton maturity - caustic soda swelling method

Around 100 fibres from Baer sorter combs are spread across the glass slide (maturity slide), and the overlapping fibres are again separated with the help of a teasing needle. The free ends of the fibres are then held in the clamp on the second strip of the maturity slide which is adjustable to keep the fibres stretched to the desired extent. The fibres are then irrigated with 18% caustic soda solution and covered with a suitable slip. The slide is then placed on the microscope and examined. Fibres are classed into the following three categories:

Mature: Rod like fibres with no convolution and no continuous lumen are classified as 'mature'.

Half-mature: The intermediate ones are classified as 'half mature'.

Immature or dead: Convoluted fibres with wall thickness one-fifth or less of the maximum ribbon width are classified as 'Dead'.

A combined index known as maturity ratio is used to express the results.

Maturity ratio = ((Mature - Dead)/200) + 0.70

About four to eight slides are prepared from each sample and examined. The results are presented as percentage of mature, half-mature and immature fibres in a sample. The results may also be expressed in terms of 'maturity coefficient'.

Maturity coefficient = (M + 0.6 H + 0.4 I)/100

where.

M is percentage of mature fibres,

H is percentage of half-mature fibres,

I is percentage of immature fibres.

Fibre maturity coefficient: Cotton fibre is classified as matured, half-mature and immature fibres. Using this classification maturity coefficient is determined. Classification of maturity coefficient is as follows.

Table 2.4 Classification of fibre based on maturity coefficient

Category	Maturity coefficient
Very immature	Less than 0.60
Immature	0.60 to 0.70
Average mature	0.71 to 0.80
Good mature	0.81 to 0.90
Very high mature	More than 0.90

2.4   Fibre strength measurement

The different measures available for reporting fibre strength are

- Breaking strength
- Tensile strength and
- Tenacity or intrinsic strength

Coarse cottons generally give higher values for fibre strength than finer ones. In order to compare the strengths of two cottons differing in fineness, it is necessary to eliminate the effect of the difference in cross-sectional area by dividing the observed fibre strength by the fibre weight per unit length. The value so obtained is known as 'intrinsic strength or tenacity'. Tenacity is found to be better related to spinning than the breaking strength.

Tensile testing

The following are the terminologies typically used in tensile testing, and their definitions are also provided as follows:

Load: The application of a load to a specimen in its axial direction causes a tension to develop in the specimen. The load is usually expressed in *grams* or *pounds*.

Breaking load/breaking strength: This is the load at which the specimen breaks. It is usually expressed in *grams* or *pounds*.

Stress: It is the ratio between the force and the area of cross-section of the specimen.

Stress = Force applied / Area of cross-section

Specific/mass stress: In case of textile material the linear density is used instead of the cross-sectional area. It also allows the strength of yarns of different linear densities to be compared.

Specific stress = Force/Linear density (initial)

The preferred units are *N/tex* or *mN/tex*; other units which are found in the industry are *gf/denier* and *cN/dtex*.

Tenacity or specific strength: The tenacity of material is the mass stress at break. It is defined as the specific stress corresponding with the maximum force on a force/extension curve. The nominal denier or tex of the yarn or fibre is the figure used in the calculation; no allowance is made for any thinning of the specimen as it elongates. Units are grams/denier or grams/tex.

Breaking length: Breaking length is an older measure of tenacity. It is the theoretical length (in Km) of a specimen of yarn whose weight would exert a force sufficient to break the specimen. It is usually measured in kilometres, for example, 10 tex yarn breaks at a load of 150 gm.

Breaking length would be = 15 km (RKm)

The numerical value is equal to tenacity in g/tex (150/10).

Strain: When a load is applied to a specimen, a certain amount of stretching takes place. The elongation that a specimen undergoes is proportional to its initial length. Strain expresses the elongation as a fraction of the original length, that is,

Strain = Elongation / Initial length

Extension percentage: This measure is the strain expressed as a percentage rather than a fraction, that is,

Extension % = (Elongation / Initial length) × 100

Breaking extension: Breaking extension is the extension percentage at the breaking point.

Gauge length: The gauge length is the original length of that portion of the specimen over which the strain or change of length is determined.

The strength characteristics can be determined either on individual fibres or on bundle of fibres.

Single fibre strength

The tenacity of fibre is dependent upon the following factors:

- Chain length of molecules in the fibre
- Orientation of molecules

- Size of the crystallites
- Distribution of the crystallites
- Gauge length used
- Rate of loading
- Type of instrument used
- Atmospheric conditions

The mean single fibre strength determined is expressed in units of 'grams/ tex'. As it is seen the unit for tenacity has the dimension of length only, this property is also expressed as the 'breaking length', which can be considered as the length of the specimen equivalent in weight to the breaking load.

Bundle fibre strength

Fibres are not used individually but in groups, such as in yarns or fabrics. Thus, bundles or groups of fibres come into play during the tensile break of yarns or fabrics. Further, the correlation between spinning performance and bundle strength is important. The testing of bundles of fibres takes less time and involves less strain than testing individual fibres. In view of these considerations, determination of breaking strength of fibre bundles has assumed greater importance than single fibre strength tests.

Cotton fibre strength - Pressley bundle strength tester

The Pressley fibre strength tester as shown in Figure 2.7 is used to test the strength of a flat bundle of fibres by gripping them between the top and bottom clamps. Cotton fibre sample is drawn at random from the bulk and is combed using coarse and fine combs respectively. These parallelized fibres are mounted in between the clamps and tightened. The protruding fringe of fibres is trimmed-off. As shown in Figure 2.6, a beam AB is pivoted at O. The rolling weight W is initially held in position by a catch and when it is lifted the catch is released and rolls down the beam. The distance traveled by the rolling weight is a measure of the load required to break the specimen. Then the clamps are removed from the tester and the two halves of the broken specimen are collected and weighed accurately. Then tensile strength is computed as follows:

1. Pressley Index (P.I.) = Breaking load in pounds / Bundle weight in mg
2. Tensile Strength = $[(10.8116 \times \text{P.I.})-0.12] \times 10^3$ pounds per square inch

$$= 5.36 \times \text{P.I. grams per tex}$$

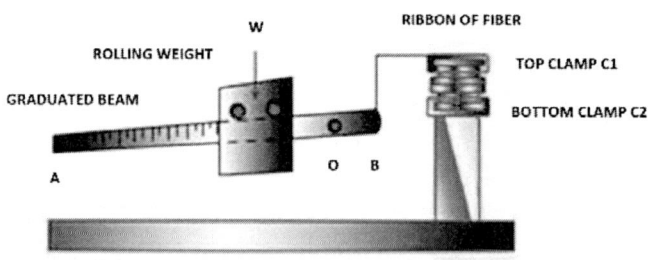

Figure 2.6 Schematic diagram of Pressley fibre strength tester

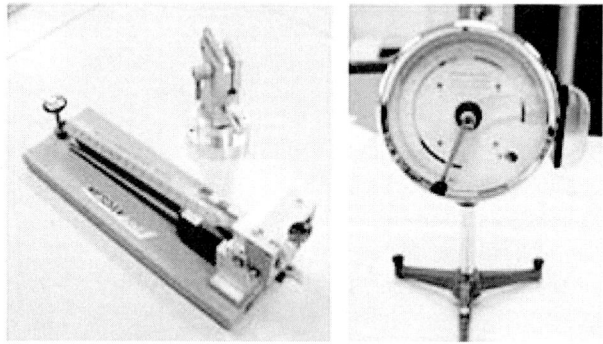

Figure 2.7 Pressley fibre strength tester

Cotton fibre strength – Stelometer

In the evaluation of the quality of raw fibre, tenacity and elongation are the two critically important physical properties to be considered. They are also important for yarn manufacturers.

The Stelometer is an instrument that tests a flat sample of cotton fibre for strength and elongation. A fibre clamp is used to hold the sample of fibre in the Stelometer. The fibre clamp holds approximately a 1/8 inch of cotton fibre. The instrument breaks the flat bundle of fibre and indicates the force required to break the fibre on a graduated scale in kilopascals. It also determines the elongation on another graduated scale at the breaking point of the fibre sample. A simple calculation is used to determine the tenacity or strength of the sample using the breaking-point number (which is indicated on the graduated scale) and the known weight of the sample. A precision balance is needed in conjunction with the Stelometer to get the weight of the sample.

Tenacity is reported in grams of force per tex unit. Tex is the number of grams a 1000-metre length weighs. Elongation is how far the fibres stretch before they break, and it is measured as a percentage of the total amount of fibre within the 1/8 of an inch (e.g., a 10% on the elongation scale; this means that 10% of the 1/8-inch fibres stretched and the other 90% did not. For both tenacity and elongation, the higher the number, the better is the cotton. If the cotton tested is found to have high tenacity, then it is stronger (e.g., Pima cotton fibres are stronger than upland varieties. A higher elongation equals a higher amount of fibre sample being stretched, which will give better results for tenacity.

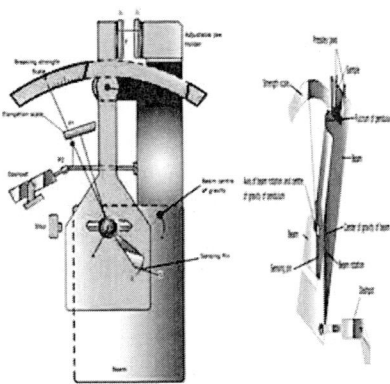

Figure 2.8 The Stelometer

A sample is loaded on the top part of the loading arm. As the trigger is depressed, the loading arm pivots to the right at the pendulum if the Stelometer is viewed from the front. The rate at which it pivots is determined by the dashpot adjustment. The pendulum is pivoted in such a way from the loading arm that its motion at first is gradual, but as it rotates further the range of motion exceeds until it stops against a rubber stop.

As it rotates, the sample in the top part of the arm is held tightly in a clamp. The clamp comes apart as the loading arm rotates stretching the fibres until they break. As they are being stretched, the force indicator and elongation indicator move along with the clamp until the fibres break. At this point the indicators stop, leaving them on the scale at a particular number.

Tensile strength / Tenacity of the fibre (in g/tex)

= Breaking load in kg x Length of sample in mm / Mass of the fibres in mg

Fibre tenacity: When fibre strength is determined without keeping distance in fibre strength tester or forceps, it is known as 0" gauge and if there is 3.2 mm distance then it is 1/8" gauge fibre strength.

Table 2.5 Classification of fibre based on tenacity

Category	Fibre tenacity (gram/Tex)	P.S.I.
Very low	Less than 34.5	Less than 6.5
Low	34.5 to 37.4	6.5-6.9
Average	37.5 to 43.0	7.0-7.9
Good	43.1 to 47.7	8.0-8.9
Very good	More than 47.5	More than 9.0

Source: CICR: Central Institute for Cotton Research

2.5   Determination of trash and lint in cotton

The determination of lint and trash content of raw cotton is important since the presence of trash directly influences the net amount of yarn or fabric that can be manufactured from a given lot of cotton. The amount of trash remaining in various intermediate products like scutcher lap, card sliver and so on indicates the cleaning efficiency of the processes or machines. Also the amount of useful lint present in the waste removed at various machines helps in making the adjustment and settings of various cleaning points of machines. Thus, the analysis of intermediate products and wastes for lint and trash contents helps in profitable adjustment and operation of the machines to clean the cotton to a predetermined degree.

Figure 2.9 Shirley trash analyser

The Shirley analyser separates lint and trash by making use of the difference in their buoyancies in the air. The specimen is fed to the taker-in cylinder with the help of feed roller and feed plate arrangement. The fibres are opened by the taker-in cylinder and are carried by an air stream and deposited on a cage similar to a condensing screen. The air stream is so adjusted that it carries only the cotton fibres and dust, leaving the trash to fall in the lower portion of the machine. The dust passes through the cage to the exhaust, and the fibres are collected in the delivery box.

Before using the machine, the delivery box, trash tray, settling chamber and so on should be swept clean. If the machine has not already been used during the day, start the motor and run the machine for 2 or 3 minutes for warming up, keeping the clutch disengaged and the feed roller inoperative.

The weight of the specimen should normally be 100 gm. Spread the specimen uniformly to cover the whole area between the guides on the feed plate, teasing out hard lumps where necessary. When making tests on slivers, short lengths should be spread on the feed plate perpendicular to the feed roller. Open the valve to its fullest extent, engage the clutch and observe the character of the trash as it begins to fall into the tray.

Only small amounts of unopened lint should be falling with the trash during the first passage, and for hard cotton it may occasionally be necessary to tighten the loading springs on the feed rollers. When the entire specimen has passed under the feed roller, as indicated by the absence of fibres under the streamer plate, disengage the clutch and close the valve momentarily to allow the lint to be collected from the delivery box.

Procedure

- Make necessary preliminary adjustments.

- Shake the specimen so that large particles of trash (which may otherwise damage the machine) are removed from the specimen; preserve these droppings for incorporation in the trash bin.

- Spread the specimen on the feed plate between the guide plates in the form of an even layer after opening out the hard lumps, if any.

- Start the machine and let the trash and lint collect in their respective compartments.

- Take out the lint from the delivery box and pass it again through the machine without disturbing the trash in the settling chamber. Stop the machine and collect the lint and keep it in a separate container (L_1).

- Remove all the trash particles containing lint from the trash tray and settling chamber and pass them through the machine. Collect the lint from the delivery box.

- Pass the lint collected as done before through the machine without disturbing the trash. Collect the lint and keep it in a separate container (L$_2$).

- Collect all the trash in the trash tray, settling chamber and any seeds clinging to the wires of the taker-in cylinder and combine them. Weigh them to an accuracy of 100 mg, and if the weight is less than 10 g, weigh to an accuracy of 10 mg (T$_1$).

- Pass the particles containing lint again through the machine and ignore the trash collected. Collect the lint and keep it in a separate container and weigh to an accuracy of 10 mg *(L$_3$)*.

- Combine all the portions of the lint *(L$_1$ L$_2$ and* L$_3$) as collected above and weigh to an accuracy of 10 mg.

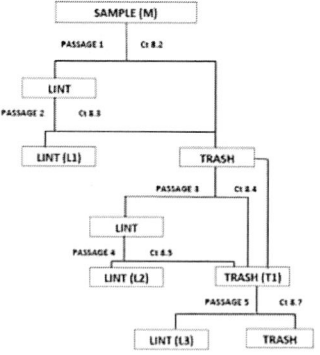

Figure 2.10 Process of trash and lint separation

Calculation

Calculate the results as lint content, trash content (visible waste content) and invisible waste content as percentages of the original specimen by the following formulae:

Lint content (L), in percentage $=[(L_1 + L_2 + L_3) / M] \times 100$

Trash content (visible waste) (T), in percentage $= [(T_1 - L_3)/ M] \times 100$

Invisible waste content (W), in percentage $= 100 - (L + T)$

where,

L_1, L_2 and L_3 = Weight of the lint portion in grams,

T_1 = Total weight of trash portion in grams,

M = Weight of the specimen in grams.

2.6   High-volume instrument (HVI)

All the methods mentioned above for the measurement of fibre properties are conventional and are mainly carried out manually. Thus, chances of getting errors are high in manual operations, and manual operations are time consuming as well. As a result, a need arose for the development of an instrument capable of measuring all properties in minimum time for better classification of cotton. HVI is the most advanced instrument for testing of fibre properties. It measures 2.5% and 50% span length, strength, elongation, micronaire, maturity ratio, percentage of maturity, fineness, UV status and colour Rd+b. It also provides uniformity ratio, short fibre percentage, mean length, upper-half mean length, uniformity index and fibrogram. HVI is patented by USTER Technologies.

Principle of HVI

HVI systems are based on the fibre bundle testing; that is, many fibres are checked at the same time and their average values are determined. Traditional testing using micronaire, Pressley, stelometer and fibrograph are designed to determine average values for a large number of fibres, which are usually called fibre bundle tests.

USTER® HVI 1000

The USTER® HVI testing system uses the latest measurement technology for the testing of large quantities of cotton samples within a short time. It is a high-performance system that permits the annual classification of entire cotton crops.

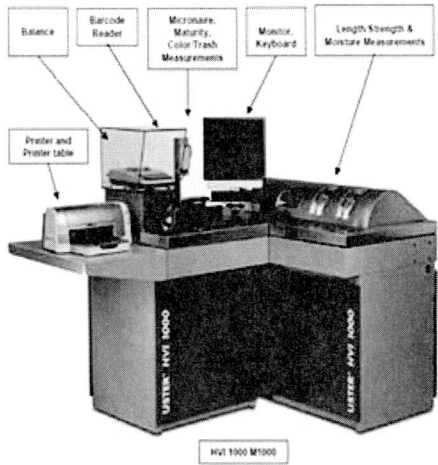

Figure 2.11 USTER high-volume instrument

Application range: 100% cotton samples in the form of bale or opened and cleaned material (card mat).

Measurements

- Micronaire
- Maturity index
- UHML - upper-half mean length
- UI - uniformity index
- SFI - short fibre index
- Fibre strength in g/tex
- Elongation
- Moisture content
- Colour (reflectance = Rd, yellowness = +b) and colour grade (USDA Upland, Pima or regional customized colour chart)
- Trash (% area, trash count) and trash grade (USDA)

Table 2.6 Measurements and calculations of USTER HVI 1000

Fibre property	Measurement / Calculation
Micronaire Reading	Measured by relating airflow resistance to the specific surface o fibres
Maturity Ratio	Calculated using a sophisticated algorithm based on several HVI measurements.
Length – Upper half mean length, Uniformity index, Short fibre index	Measured optically in a tapered fibre beard which is automatically prepared, carded, and brushed.
Strength and Elongation	Strength is measured physically by clamping a fibre bundle between 2 pairs of clamps at known distance. The second pair of clamps pulls away from the first pair at a constant speed until the fibre bundl breaks. The distance it travels, extending the fibre bundle before breakage, is reported as elongation.
Moisture content	It is measured using conductive moisture probe.
Colour – Rd(whiteness), +b (yellowness), colour grade	Measured optically by different colour filters, converted to USD Upland or Pima colour grades or regional customized colour chart.
Trash – particle count, % surface area covered by trash, Trash code	Measured optically by utilizing a digital camera, and converted to USDA trash grades or customized regional trash standards

Additional features include the following:

- Safety interlocks to prevent injury from unauthorized entry to the instrument

- Relative humidity and temperature probe
- Moisture measurement
- Easily accessible lint and waste box with two separate access doors
- Computer system can be easily removed for service
- Configuration can be straight-line configuration or 'L' configuration
- Industrial brushed stainless steel top and work surfaces
- Single-point adjustable brush pressure
- Integrated air enclosure around balance to eliminate influences of air turbulence
- Password-protected operational software.

Source: www.uster.com

3.1   Yarn numbering system - count

Count is a numerical value which expresses the coarseness or fineness (diameter) of the yarn and also indicates the relationship between length and weight (the mass per unit length or the length per unit mass) of that yarn.

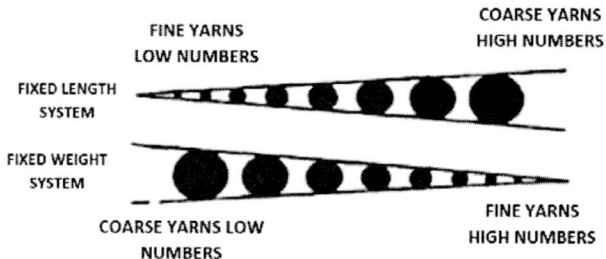

Figure 3.1 Yarn numbering principle

The fineness of the yarn is usually expressed in terms of its linear density or count. There are a number of systems and units for expressing yarn fineness. They are classified as follows:

There are two types of yarn count system:

- Direct count system
- Indirect count system

Direct count system

The weight of a fixed length of yarn is determined. The weight per unit length is the yarn count. The common feature of all direct count systems is the length of yarn fixed, and the weight of yarn varies according to its fineness. The following formula is used to calculate the yarn count:

$$N= W / L$$

where,

N = Yarn count or numbering system

W = Weight of the sample at the official regain in the unit of the system

L= Length of the sample

Table 3.1 Direct yarn count system

Numbering System	Unit of Length (l)	Unit of Weight (w)
Tex system, T	1000 metres	No. of Grams
Denier, D	9000 metres	No. of Grams
Decitex, dtex	10000 metres	No. of Grams
Millitex, mtex	1000 metres	No. of Milligrams
Kilotex, ktex	1000 metres	No. of Kilograms
Jute count	14400 yards	No. of Pounds (lb)

a. Denier system

Weight in grams of yarn/fibre of 9000 metre length is called the Denier of that particular yarn/fibre.

Denier = weight in grams of 9000 metres length

b. Tex System

Weight in grams of yarn/fibre of 1000 metre length is called the Tex of that particular yarn/fibre.

Tex = weight in grams of 1000 metres length

Indirect count system

The length of a fixed weight of yarn is measured. The length per unit weight is the yarn count. The common feature of all indirect count systems is the weight of yarn fixed, and the length of yarn varies according to its fineness. The following formula is used to calculate the count:

$N = L / W$

where,

N = Yarn count or numbering system

W = Weight of the sample at the official regain in the unit of the system

L = Length of the sample.

Table 3.2 Indirect yarn count system

Numbering System	Unit of Length (l)	Unit of Weight (w)
English cotton count, N_e	840 yards (yds)	1 pound (lb)
Metric count, N_m	1000 metres / 1 km	1 kg
Woollen count (YSW)	256 yards	1 pound (lb)
Woollen count (Dewsbury)	1 yard	1 ounce (oz)
Worsted count, NeK	560 yards	1 pound (lb)
Linen count, NeL	300 yards	1 pound (lb)

1. N_e: No. of 840 yards yam weighing 1 pound

2. N_m: No. of 1 km yarn weighing 1 kg

The Ne indicates how many hanks of 840 yards length weigh one English pound; so, 32 Ne Means 32 hanks of 840 yards, that is, 32 x 840 yards length weigh 1 pound. For the determination of the count of yarn, it is necessary to determine the weight of a known length of the yarn. For taking out known lengths of yarns, a wrap-reel is used. The length of yarn reeled off depends on the count system used. One of the most important requirements for a spinner is to maintain the average count and count variation within control.

Conversion of count

The conversion of yarn count from one system to another may be made easy with the following conversion table.

Table 3.3 Conversion of yarn count from one system to another

Given / Determined	N_e	N_m	Tex	Decitex	Denier
N_e	1.0	$N_m/1.69$	590.5/Tex	5905/dtex	5315/D
N_m	$1.69 \times N_e$	1.0	1000/Tex	10000/dtex	9000/D
Tex	590.5/ N_e	1000/N_m	1.0	Decitex × 0.1	Denier/9
Decitex	5905/ N_e	10000/N_m	Tex × 10	1.0	Denier/0.9
Denier	5315/N_e	9000/N_m	Tex × 9	Decitex × 0.9	1.0

Conversion of count - examples

1. Conversion of 30 Ne to Tex and Denier

   Using Table 3.3

   30 Ne = 590.5/30 Tex = 5315/30 Denier 30

   Ne = 19.68 Tex = 177.16 Denier

2. Conversion of 120 Ne to Tex and Denier

   120 Ne = 4.92 Tex = 44.29 Denier

3. Conversion of 200 Denier to Tex and Ne

   200 Denier = 200/9 Tex = 5315/200

   Ne 200 Denier = 22.22 Tex = 26.58 Ne

3.2   Determination of yarn count

Instruments for count determination

To determine the yarn count of a sample, it is needed to measure the length and weight of the sample. The equipments used for this purpose are wrap reel and analytical balance, Knowles balance, Quadrant balance and so on. If the yarn specimen supplied is not sufficient to perform the tests using the above methods, then Beesley balance can be used to examine the yarn count with reliability. Beesley balance can be used to get the yarn count directly from the balance.

Quadrant balance can be used to measure the count of yarn containing the length less than 120 yards, and this gives a direct reading of yarn count. It consists of a quadrant scale fixed to a pillar as shown in the figure 3.2. At the top of the pillar, pointer is pivoted so that it moves over the face of the quadrant scale.

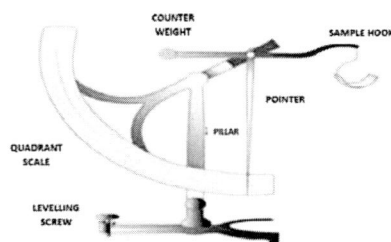

Figure 3.2 Quadrant balance

A cross-beam is also pivoted in the same pivot of the pointer which has a sample hook at one end and a counter weight at the other end. The counter weight determines the capacity of the instrument. Addition of any weight to the sample hook makes the pointer to move in front of the scale. The quadrant scale is divided into three scales. The top scale can be used to find the weight per square yard of a cloth sample in ounces, the middle scale to find the count of yarn 8 yards long and the third scale to find the count of yarn 40 yards long.

On the top of the pointer and the beam pivot, there is a small adjusting screw, the adjustment of which brings the pointer in line with the datum line. The instrument can also be levelled with a levelling screw provided at the base of the instrument.

To operate the instrument, it is calibrated after levelling with the base screw. A counter weight marked 40^s is used for calibration. For this, the counter weight is suspended from the sample hook. If the balance is on level, the pointer reads 40ˢ on the 40-yard scale. If it does not read 40ˢ, the pivot of the pointer is adjusted until it reads 40ˢ on the 40-yard scale.

If a sample of 8 yards length is used, its count can be noted from the 8-yard scale and if a sample of 40 yards length is used its count can be noted from the 40-yard scale.

Beesley balance

Beesley balance works on the principle of fixed weight and fixed length system. It is used when the warp and the weft count of yarn needs to be measured for a small piece of fabric. The fabric is cut into small lengths with a template. Yarns are then removed from the specimen, and the total number of yarn lengths required to balance a standard weight on the beam directly gives the count of yarn.

Beesley balance consists of a light-weight beam pivoted on jewel bearings with a hook at one end and a pointer at the other end. The beam is initially levelled to bring the pointer against a datum line. A standard weight is suspended in a notch on the beam arm on the pointer side.

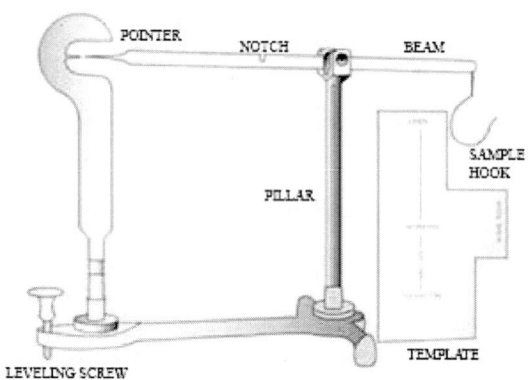

Figure 3.3 Beesley balance

A template is used to cut short lengths of yarn, with the length depending on the count system required. These short lengths are placed on the hook until the pointer comes against the datum line. The number of short lengths required to balance the beam gives the count of the yarn. A template is provided to give lengths of yarn in metric, cotton, linen, wool skein and worsted counts. The balance is housed in a sheet metal box having sliding transparent acrylic doors.

3.3   Determination of yarn strength

As yarn forms the basis of production of all types of fabrics it is essential that the strength and elongation of the yarn is monitored by a yarn strength testing system to ensure that it is suitable for today's high-speed production techniques and its performance matches the requirements of the finished product.

- *Tensile*: The tension capacity in a material or tension exerted on a material.

- *Strength:* The property of a material that resists deformation induced by external force.

- *Tensile strength:* The strength of a material under tension which is distinct from compression, fusion or shear.

Principles of CRT, CRL and CRE

There are three ways to carry out tensile test:

1. Constant rate of extension (CRE): Here, the rate of elongation of the test specimen is kept constant and the load is applied accordingly.

2. Constant rate of loading (CRL): Here, the rate of increase of the load is kept uniform with time, and the rate of extension is dependent on the load elongation characteristics of the specimen.

3. Constant rate of traverse (CRT): Here, one of the clamps is pulled at a uniform rate and the load is applied through the other clamp that moves to actuate a load-measuring mechanism so that the rate of increase of either load or elongation is usually not constant.

CRE versus CRL

- With CRE principle, maximum load is reached before 3 seconds, and the rest of the time the specimen remains at a higher load (initially a very high rate of loading).

- For CRL, initial extension is low and afterwards a very high extension occurs within a small duration (very high extension rate)

- If specimen length increases in a CRE machine, the rate of loading will decrease.

Single yarn strength

The single yarn strength tester which works on the basis of the pendulum lever principle is shown in Figure 3.4. It is a motor-driven instrument. It is made up of a weight arm pivoted in ball bearings and has a quadrant at the top. The arm is connected to an upper clamp by a chain which runs over the quadrant and fixed to it. The pulling force acting on the specimen is transferred to the pendulum through the clamping arrangement which in turn displaces the weight in proportion to its own magnitude, and this can be read on the quadrant scale.

This is made in two ranges: one for the lower range and the other for higher range. The lower scale is used when there is a small weight on the pendulum, and the higher scale is used when the pendulum carries additional weight.

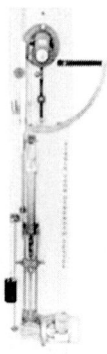

Figure 3.4 Single yarn strength tester

The lower jaw is carried by a rack which is connected to a driving mechanism consisting of a clutch and declutch arrangement. A rod which carries the lower jaw can be changed to different length so that specimens with different lengths can be tested. The specimen will break within 20 to 23 seconds. The rate of traverse of the bottom jaw is 300 ± 15 mm/minute.

Testing: The pendulum is arrested with a catch and also the movement of the upper clamp is arrested. The material to be tested is taken and clamped between the upper clamp and the lower clamp. The extra material is cut off exactly at the clamp position and then the catches are taken out. When the machine is started, the lower jaw traverses downward imposing the tension on the specimen and thereby pulling the upper clamp and in turn this makes the pendulum to move over the quadrant scale. When the specimen is ruptured, the pendulum arm is retained in the position by a set of pawl working over the serrated portion of the quadrant. The position of pendulum arm gives the breaking load of the specimen. Comparatively, 50 tests are done for single yarns and 25 tests for plied yarns, every time bringing the pendulum arm to zero position and arresting the movement of the upper jaw. Then, the tenacity is calculated using the formula,

Tenacity in g per tex = Mean breaking load in kg × 1000 / Tex

Apart from the breaking load, the elongation is also measured by noting down the relative position of the upper clamp. The elongation scale directly reads the difference in the movement of both the lower and upper jaws. This can also be calculated in terms of percentage of the original length of specimen as follows:

Elongation % = (Elongation scale reading / gauge length) × 100

Lea strength

The lea tester shown in Figure 3.5 works based on the CRE principle, in particular the pendulum lever type.

Figure 3.5 Lea strength tester

Description: It is a motor-driven, pendulum-type strength tester. It consists of an upper jaw and a lower jaw. The lower jaw can be engaged with a screw mechanism, which is driven at a constant speed by a motor. Thus, a constant rate of traverse of 12 inches/minute is given to the bottom jaw. The top jaw is connected to a pendulum arm by means of a steel tape. A heavy bob is attached to the pendulum arm, and the arm moves over a serrated quadrant. A pawl is attached to the pendulum to control the movement of the arm and helps the arm to stop when the lea breaks. There is a dial, calibrated in pounds. over which a pointer moves through the geared movement of the pendulum arm. The pointer moves through the geared movement of the pendulum arm. The pointer indicates the lea strength in pounds on the dial.

Testing: Samples of lea are prepared from the ring bobbins or cones using wrap reel. The length of a lea is 120 yards. The bobbins are reeled under the same tension and with a small traverse to separate the layers. When the set length is wound, the reel automatically stops and the lea is transferred from the wrap reel to the lea tester. The lea is mounted over the jaws and while doing so care should be taken to avoid the formation of any twist in the lea and the groping of threads on the jaws. Then the bottom jaw is engaged with the screw mechanism and the motor is switched on.

As the bottom jaw descends, a load is imposed on the loops of yam constituting the lea. As a result of the pull on the upper jaw the pendulum arm is pulled and the pointer moves over the dial as a result. At one point, one or two strands may break and many will slip too; at that stage, the pendulum will stop moving. The pendulum will be prevented from falling back suddenly by the pawl which engages with the teeth over the serrated quadrant. At that place, the pointer also stops moving and indicates the maximum load on the dial. This load is called the strength of the lea. The lower jaw is then brought up and the lea is removed from the jaws.

Universal tensile strength tester

The single yarn and lea strength testers mentioned above are mechanical-type equipments. Now, with the advent of computerised universal tensile strength tester all kinds of tensile testing can be done with a single machine just by changing the sample mounting clamps. Tests such as single yarn strength; lea skein strength; fabric tensile tests such as strip test, grab test and so on; peel bond strength; button/snap pull strength; seam slippage; and zipper strength can be carried out using universal tensile strength tester. It is very fast and automatic. The test results are transferred to a computer, and all data (test results) can be stored and also printed whenever needed.

Determination of yarn count and count strength product (CSP)

When yarn is available in package form, yarn count is tested using wrap reel and weighing balance. The CSP is a useful measure of yarn strength and it helps in making comparisons among yarns of similar count.

Sample preparation

Yarn samples from at least five packages are drawn. From each package skeins of lea are prepared using wrap reel. Prior to sample preparation same part of yarn is unravelled from the package and discarded; in the same way, after preparing one sample from each package, some amount of yarn is discarded and the second lea is prepared.

Procedure

- Prepare 10 leas of yarn from the given yarn packages (sample)
- Weigh each lea in a precision balance and tabulate the results. Find the mean and calculate the count using the formula, Count, Ne = 64.78 / mean weight of lea in grams
- If direct yam count balance is available, it may be used to find the count directly.
- Now take the leas to the tensile tester and find out the tensile strength of all the lea samples. Tabulate the result.
- Calculate the CSP as follows:

    CSP = Yam count (Ne) x Lea strength (pounds)
- Find the average and report the results.

3.4   Determination of yarn twist

Twist is the measure of the spiral turns given to yarn in order to hold the fibres or threads together. Twist is necessary to give coherence and strength to the yarn. Twist is primarily inserted into a staple yarn to hold the constituent fibres together, thus giving strength to the yarn. False twist is used in textured yarns.

The effects of the twist are twofold:

1. As the twist increases, the lateral force holding the fibres together is increased so that more fibres could be contributed to the overall strength of the yarn.

2. As the twist increases, the angle that the fibres make with the yarn axis increases, which prevents them from developing their maximum strength, which occurs when they are oriented in the direction of the applied force. As a result, at certain point the yarn strength reaches a maximum value after which the strength is reduced as the twist is increased still further as shown in Figure 3.6(a).

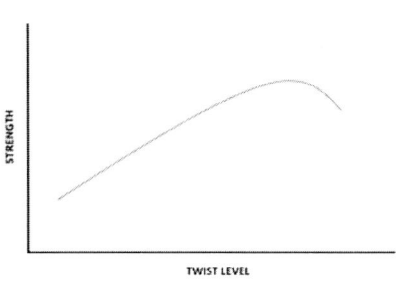

THE EFFECT OF TWIST LEVEL ON STRENGTH, STABLE FIBER YARN

Figure 3.6(a) Effect of twist level on strength of staple fibre yarn

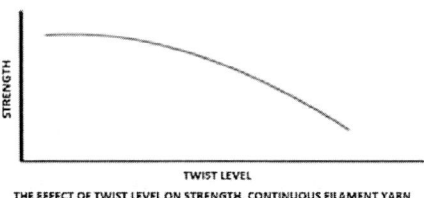

THE EFFECT OF TWIST LEVEL ON STRENGTH, CONTINUOUS FILAMENT YARN

Figure 3.6(b) Effect of twist level on strength of continuous filament yarn

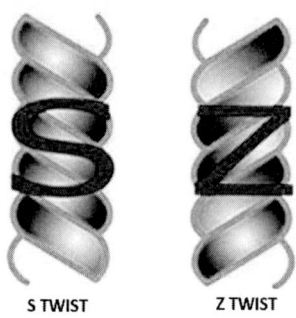

S TWIST   Z TWIST

Figure 3.7 Direction of twist in spun yarn

Direction of twist

S-twist: The twist in yam due to which its spirals are in line with the central portion of the letter S, when the yarn is held in a vertical position.

Z-twist: The twist in yarn due to which its spirals are in line with the central portion of the letter Z, when the yarn is held in a vertical position.

Twist multipliers

Relationship between yarn count and twist

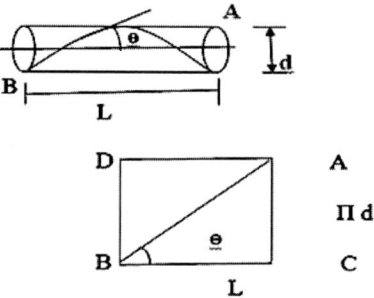

Figure 3.8 Schematic diagram of twist in spun yarn

From Figure 3.8, it can be seen that

$$\tan \theta = \pi\, d\, /\, L$$

Where θ = twist angle, d = yarn diameter and L= yarn length.

Also from Figure 3.8, it can be seen that the height (pitch) of one turn of twist is L. Since the twist level is normally specified as the number of turns per metre, the twist level in 1 metre of the yarn would be

twist = 1/L

Thus, tan θ = n d x twist

Also, it is evident from experience that yarn diameter is very hard to measure because textile yarns by their very nature are soft and squashy. On the other hand, yarn count is normally used to express the thickness of the yam. But we can relate yarn diameter to yarn count using the following expression:

Cubic density (p) = linear density (Tex) / cross-sectional area (A)

Assuming a circular cross-section for the yarn, we get

ρ **(g/m³) = Tex × 10⁻³ (g/m) / A (m²) = Tex × 10⁻³/ πd²/4**

Solving for *d*,

d² = 4 Tex x 10⁻³ / ρ π

Twist = tan 0 / 2n Î tex x 10⁻³ / ρ π

= K / $\sqrt{}$tex where K = 0.5 tan 0 $\sqrt{}$ ρ 10³ / π

K is called the twist factor and is proportional to 0 if p remains constant.

Thus, K is a factor relating twist level to yarn count. The derivation shows that if two yarns have the same twist factor, they will have the same surface twist angle, regardless of count. Since surface twist angle is the main factor determining yarn character, twist factor can be used to define the character of a yarn. It is worth noting there can be minor errors associated with the use of twist factor for the following reasons:

1. The cubic density may be different for different yarns. It is assumed in the foregoing calculation that this will not change for yarns of the same surface twist angle.

2. Different fibres with different frictional and other properties will create different yarn characters.

Nevertheless, the relationship derived between twist, twist factor and yarn count is one of the most important in the study of yarn technology. This relationship is expressed in different ways for different yarn count systems.

For the tex system:

Twist (turns per metre) = Twist Factor (K_t) / $\sqrt{}$ tex

For the English count (Ne) system:

Twist (turns per inch) = Twist Factor (K$_e$) $\sqrt{}$ Ne

Twist tester - tension type

The tension-type twist tester is working on the principle of twist contraction. This method is also called untwist-re-twist method. Suppose a yarn is twisted in Z direction and has a length L. Let the twist be completely removed to produce an untwisted strand of length L + C where C is the contraction due to twist. If the strand is now twisted in S way with a number of turns equal to those removed, it can be expected that the strand will again contract to the original length L. This method is suitable for single yams.

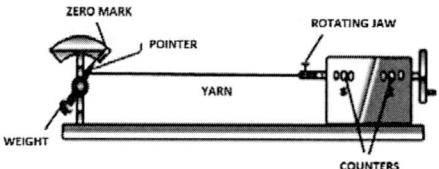

Figure 3.9 Tension-type twist tester

The tension-type twist tester as shown above has two jaws (fixed and rotating) for fixing the yarn. The position of the non-rotating jaw is variable according to the gauge length. The pointer connected to the rotating jaw moves over a scale having two graduations.

The rotating jaw shaft has a worm at the middle to give drive to the mechanical counter. A pinion is provided at the other end of the shaft and is rotated by the hand wheel through the gear. As the handle rotates, the rotating jaw rotates and the rotations are transferred to the mechanical counter which consists of two discs, one at the bottom and one at the top. The top disc is graduated into 100 divisions and has 100 teeth. Placed over the top disc is a pointer which is connected to the bottom disc that has 99 teeth. Therefore, for one revolution of the top disc, the pointer moves only 99 teeth and always lags one tooth. This helps in determining the dial reading more than 100.

The rotating jaw assembly has a spring-loaded knob. By pushing this, the two discs of the counter can be disengaged from the worm for zero setting. There is an index mark on the frame of the rotating jaw. For zero setting, the pointer and the zero mark of the counter should coincide with the mark on the frame.

Procedure

The two discs are disengaged from the worm and are rotated such that the pointer and zero mark coincide with the index mark. The yarn is first gripped in the fixed clamp. After being led through the rotating jaw, the yarn is pulled through until the pointer lies opposite a zero line on a small quadrant scale. The jaw is then closed. At this stage, the specimen is under a small tension and has a nominal length (gauge length).

As the twist is removed the yarn gets extended and the pointer reaches a vertical position. Eventually all the twist is taken out, but the jaw is kept rotating in the same direction until sufficient twist has been inserted to bring the pointer back to the zero mark. When the pointer coincides with the mark, the dial reading is noted and the twists per inch can be calculated using the following formula:

Twist per inch = dial reading / (gauge length × 2)

Automatic/motor-driven twist tester

Figure 3.10 Automatic/motor-driven twist tester

Sample preparation

- From each packages, an equal number of specimens of adequate length shall be selected from different parts of the package.

- The exact length may be either 350 or 500 mm, and the total quantity of test specimens is 10% of the test samples available.

- Before cutting each test specimen from the package, a length of about 10 m of yarn should be discarded.

- While drawing the specimens from the packages, care should be taken to avoid any changes in the twist of yarn.

Procedure

- The yarn specimen is placed on the yarn stand; keep the yarn stand on the left-hand side of the instrument.

- Draw the yarn through the yarn guide which is in straight-line to the yarn bobbin, and draw through the instrument's yarn guide, which is fixed at the tail end on the left side of the instrument.

- Depress the center screw of the disc yarn gripper using the thumb and lift the outer cone by force and middle finger of the left hand; then catch the yarn by right hand, drawing through the opening of the disc yarn gripper and pull the yarn to required length (i.e., slightly more than the gauge length). Then arrest the yarn by simply releasing the outer cone.

- After applying correct tension, release the hand from the revolving shaft gently and cut the protruding yarn by a sharp knife. Now, the specimen yarn is ready for test.

- Press the 'Start' button. Regulate the motor RPM to set the appropriate speed. While setting motor speed, the tester should ensure that the following parameters are followed:

1. Set the motor speed at half-speed of maximum speed at which the motor can rotate; this is specifically applicable to all cotton yarns.

2. Set the motor speed at 75% speed for high-twist filament yarns.

Testing of single yarn

For single yarns the test will stop automatically after the completion of untwist – retwist procedure. The display will show the direct result in TPI or TPM.

Testing of double and plied yarns

For double and plied yarns only the untwist process is used. In this method test will stop automatically at a pre-set value. The pre-set value may be 2 to 3 less than the calculated TPI. Now insert a sharp-edged needle between the opened double yarn at the extreme end of the specimen yarn (from disk yarn gripper) and move towards the rotating jaw and press the 'Start' button once for releasing the remaining twist in the yarn. Reduce the motor's speed to minimum. By doing this the motor will stop exactly after releasing the twist. The display will show the direct result in TPI.

3.5   Measurement of yarn evenness

Regardless of the different terms used, such as evenness or unevenness and regularity or irregularity, they ultimately denote the degree of uniformity of a product. For textile products such as laps, slivers, rovings, and yarns, which are the products of various spinning machines, the level of uniformity is expressed in terms of *evenness* or *regularity* or in terms of *unevenness* or *irregularity.*

Producing a yarn with uniform characteristics such as uniformity in weight per unit length, uniformity in diameter, turns per inch and strength is indeed complex, and perfect uniformity is at best a dream. This is more because in the case of staple fibres, especially in natural fibres, transforming millions of individual fibres of varying fineness, maturity, length, colour, diameter and so on into a yarn of uniform character is at best hypothetical and thus impossible. Another important aspect about natural fibres is there is a lot of variation in length, colour and thickness, which leads to variations in yarns and fabrics.

Causes of irregularity

1. Properties of raw material
2. Fibre arrangement in the yarn
3. Fibre behaviour
4. Inherent shortcoming of machinery
5. Mechanically defective machinery
6. External factors such as working condition and inefficient operation

Classification of variation

There are two types of variation: Random variation and Periodic variation.

Random variation is the variation which occurs randomly in the textile material and can occur in any order. Suppose a yarn is cut into short, equal lengths, say, 1 inch, then the weight of each consecutive lengths can be measured as shown in Figure 3.11. The weights are plotted in a graph against the lengths similar as shown in Figure 3.11:

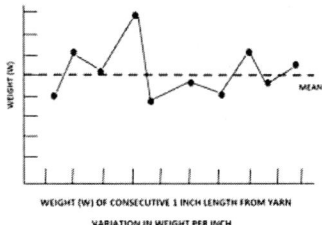

WEIGHT (W) OF CONSECUTIVE 1 INCH LENGTH FROM YARN

VARIATION IN WEIGHT PER INCH

Figure 3.11 Random variation in yarn

All traces of irregularity do not show random distribution of deviations from the mean. Suppose traces show definite sequences of thick and thin places in the strands. These forms of irregularity are called as periodic variations. Periodic variations are the variations with definite sequences of thick and thin places in the strands.

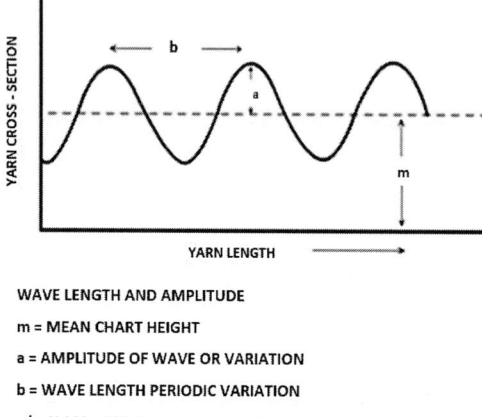

WAVE LENGTH AND AMPLITUDE

m = MEAN CHART HEIGHT

a = AMPLITUDE OF WAVE OR VARIATION

b = WAVE LENGTH PERIODIC VARIATION

a/m X 100 = PERCENTAGE AMPLITUDE

Figure 3.12 Periodic variation in yarn

There are two terms used to describe a periodic variation - *wave length* and *amplitude* - as shown in Figure 3.12. Wave length is the distance between one peak of the wave and the next on the same side of the mean line. Amplitude is a measure of the size of the swing from the mean level. Usually, this is expressed as a percentage of the mean.

Short, medium and long-term variations

Using the fibre length as a length unit, the periodic variations in the fibrous strand are classified according to their wavelength with respect to the fibre length used for forming that particular strand. The different classifications of variations are listed as follows:

1. *Short-term variation*: Wave length is 1 to 10 times the length of the fibre.

2. *Medium-term variation*: Wave length is 10 to 100 times the length of the fibre.

3. *Long-term variation:* Wave length is 100 to 1000 times the length of the fibre.

This classification is used for investigating and determining the causes of faults. The amplitude of short-term variation is generally greater than that of the long-term variation because it occurs at the last machine and has not been reduced by doubling.

Methods of measuring evenness

1. *Visual examination methods*

Black boards, drums, photographic devices, projectors and lap meter

2. *Cutting and weighing methods*

Lap scale, lap meter, sliver, roving and yarn wrapping

3. *Electronic capacitance testers*

Fielden-Walker evenness tester and Uster evenness tester

4. *Variation in thickness under compression*

WIRA roving levelness tester and LINRA roller yarn diameter tester

5. *Photoelectric testers*

WIRA photoelectric testers and LINRA tester

6. *Miscellaneous methods*

Airflow, mercury displacement and so on

Black board appearance method

Yarn to be examined is wrapped onto a matt black surface in equally spaced turns. The black boards are then examined under good lighting conditions using uniform non-directional light. ASTM has a series of Cotton Yarn Appearance Standards which are photographs of different counts with the appearances classified into four grades. The test yarn is then wound on a blackboard (measuring 9.5 x 5.5 inches in surface area) with correct spacing and compared directly with its corresponding standard.

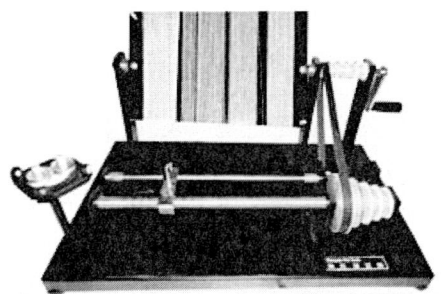

Figure 3.13 Hand-driven wrapping machine

Figure 3.14 Motor-driven wrapping machine

Both hand-driven and motorized wrapping machines are available as shown in Figures 3.13 and 3.14. The yarn is made to traverse steadily along the board as it is rotated, thus giving a more even spacing. It is preferable to use tapered boards for wrapping the yarn if periodic faults are likely to be present. This is because the yarn may have a repeating fault of spacing similar to that of one wrap of yarn. By chance it may be hidden behind the board on every turn with a parallel- side board, whereas with a tapered board it will at some point appear on the face.

ASTM Standards

ASTM standard test method classifies the yarn appearance into five grades. The board is compared with standard photographs and then graded.

- *Grade A:* No large neps, very few small neps, must have very good uniformity, less fuzziness.

- *Grade B:* No larger neps, few small neps, less than three small pieces of foreign matters per board, slightly more irregular and fuzzy than A.

- *Grade C:* Some larger neps and more smaller neps, fuzziness, foreign matters more than B, more rough appearance than B.

- *Grade D:* Some slubs (with diameter 3 times the diameter of yarn). More neps, larger neps, fuzziness, thick and thin places, foreign matters than in Grade C yarn. Overall appearance is rougher than C.

- *Grade E:* Below grade D; more defects and overall rougher appearance than grade D yarn.

Yarn appearance indices

Based on the yarn appearance grades, each grade is assigned an index value as shown in Table 3.4.

Table 3.4 Yarn appearance indices

Grade	Designation	Index
A	EXCELLENT	130
B +	VERY GOOD	120
B	GOOD	110
C +	AVERAGE	100
C	FAIR	90
D	POOR	80
D +	VERY POOR	70
BG	BELOW GRADE	60

USTER evenness tester

USTER evenness tester is used to calculate the unevenness (U%), coefficient of variation of mass (CVm%), yarn hairiness, imperfection index (IPI) and thick, thin place, neps and so on of yarn, roving and sliver.

The evenness of yarn is one of the main indices to be used to measure the quality of yarns. The unevenness of yarns will deteriorate the mightiness of yarns and increase the end-breakage rate in the spinning, and the increase of end-breakage rate will directly limit machine speed and reduce productivity. In addition, unevenness of yarns will seriously influence the quality of appearance of textiles.

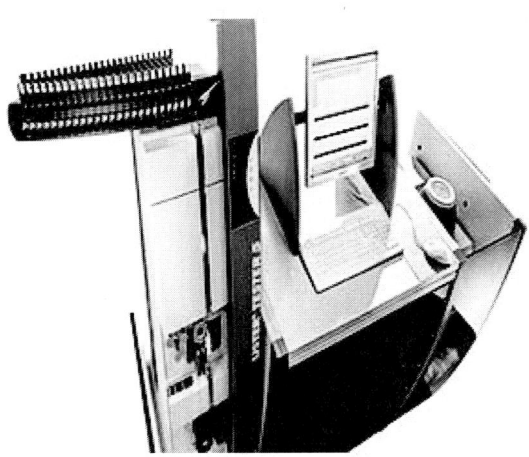

Figure 3.15 USTER evenness tester

Principle of USTER evenness tester

Raw material as well as spinning problem can be detected by measuring yarn unevenness, which can be done by using USTER evenness tester or USTER Tester-5 (UT 5). The quality parameter is determined by a capacitive sensor. In this case the yarn, roving or sliver is passed through the electric field of a measuring capacitor. Mass variation of material causes disturbance in the electric field, which is then converted into electric signal. This is proportional to the mass variation of material. The unevenness is recorded as a diagram.

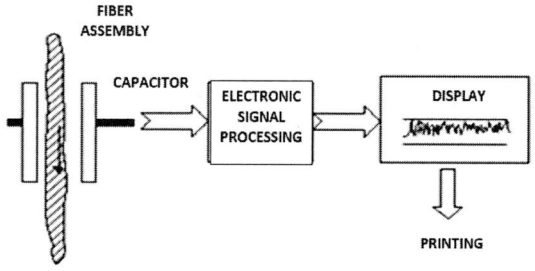

Figure 3.16 Principle of USTER evenness tester

Advantages of USTER evenness tester

- The CV percentage measured by USTER gives a measure of the variation of weight per unit length.
- This instrument measures the irregularity of material at high speed (2- 100 ft/min).
- It can show the percentage of both MD and CV of material.
- The recorder of pen can work at a high speed of 100 yards/min.

Uses of USTER evenness tester

1. Evenness measurement of yarn, roving and sliver
2. Measurement of imperfection (thick, thin place, neps)
3. Mass analysis
4. Spectrogram analysis/frequency analysis
5. Yarn hairiness measurement

6. Fabric simulation, that is, before making fabric; this way knowledge is gained about the yarn quality that ultimately determines the quality of fabric

7. Variation of trend analysis

3.6   Yarn faults classification

1. *Frequently occurring faults:* These are faults occurring in the range of 10 to 5000 times per 1000 m of yarn. Yarns spun from staple fibres contain imperfections, which can be subdivided into three groups:

a. Thin places: Cross-sectional size -30% to -60% of normal yarn with fault length ranging from 4 to 25 mm.

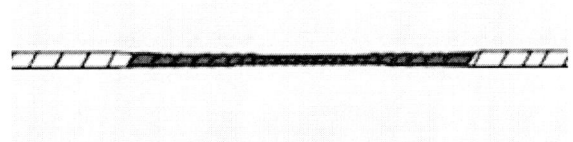

Figure 3.17 Thin places in yarn

b. Thick places: Cross-sectional size +30% to +100% of normal yarn with fault length ranging from 4 to 25 mm.

Figure 3.18 Thick places in yarn

c. *Neps:* Neps are defined as small, tight balls of entangled fibres seen in linear textile strands. Cross-sectional size +140% to +400% of normal yarn with a fault length of 1 mm.

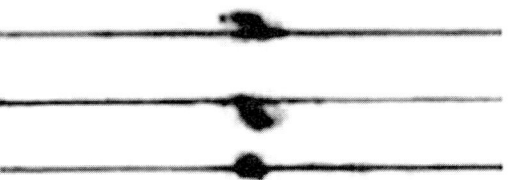

Figure 3.19 Neps in yarn

2. Seldom occurring faults: These are the thick and thin places in yarn which occur so rarely that spotting them would require testing at least 100,000 m of yarn. These faults may be classified further into the following types:

a. Short thick places: 1 to 8 cm and above +100%

b. Long thick places: Above 8 cm and above +45%

c. Long thin places: Above 8 cm and less than -30%

USTER classimat

It is a capacitor-type sensing unit and is used to count the various types of imperfections. A series of Classimat grades were arbitrarily established as shown in Figure 3.21. The yarn faults are divided into 16 grades based on their length and linear density. If a fault passes through the sensing capacitor, its length and linear density are monitored and classified according to this grading system.

Figure 3.20 USTER Classimat

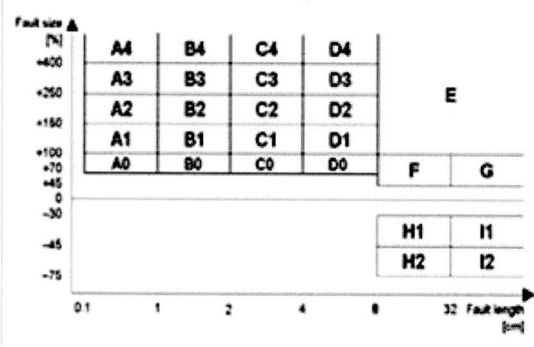

Figure 3.21 Classimat yarn faults classification

A Classimat installation consists of six sensing units which are applied to six yarns wound on a winding machine. These units are each connected to the same classifying instrument. Thus, signals from all yarns are fed into the same counters. This instrument has a display of 16 counters. They are so connected that A1 counts all faults shorter than 1 cm whose linear density is more than double the average linear density of the yarn, A2 also counts all faults shorter than 1 cm but only those with mean linear density in excess of 150% and A3 counts all faults shorter than 1 cm but having a greater mean linear density exceeding 250%. Similarly, all class B faults (1-2 cm in length) with more than double mean linear density are recorded at B1, those of class B greater than +150% at B2; that is, within each length group the classifying instrument provides cumulative frequency values. To obtain actual numbers in groups 1, 2, 3 and 4, the number in 2 should be subtracted from 1 and the number in 3 from 2.

USTER® CLASSIMAT 5 offers the most technically advanced sensors and superior hardware to detect and eventually classify all types of defects. The unique USTER® sensor range has all the options covered:

- The new capacitive sensor identifies both short and fine neps and troublesome thick and thin places that could not be detected up until the time they show up in the final fabric.

- The latest foreign fibre sensor with using multiple light sources can locate and classify contamination in yarns, even separating colour fibres and vegetable matter in cottons and blends, and distinguish potentially non-disturbing materials from real defects.

- Breakthrough technology in polypropylene detection, which is made possible with a novel combination of hi-tech sensors. A novel sensor combination enables polypropylene content to be detected and classified for the first time.

- Has a unique range of advanced sensors covering every option for yarn defect classification.

- The new capacitive sensor comes with enhanced detection capabilities.

- Foreign-matter technology can distinguish all colours and non-disturbing materials.

- New mounting module, special guides and tension control, all of these ensure best possible accuracy in results.

3.7   Measurement of Yarn hairiness

Fibres protruding out of the main body of the yarn are called hairiness. In most circumstances it is an undesirable property, giving rise to problems in fabric production. By nature, hairiness in short staple fibre yarns is the reason for wide differences in fibre thickness, maturity and inadequate spinning process. In filament yarns, hairiness occurs due to weak monofilaments, inadequate finish, rough surfaces and loose running monofilaments which break during the subsequent processes.

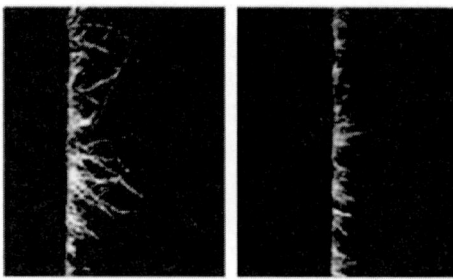

Figure 3.22 Hairiness in yarn

It is not possible to measure hairiness with a single parameter because the number of hairs and the length of hairs vary independently. A yarn may have a small number of long hairs or a large number of short hairs or any combination in between. The problem then is determining which combination should be given a higher hairiness rating. It is considered that there are two different exponential mechanisms in operation, one for hairs with a length of 3 mm and more and another for those with length shorter than 3 mm. The percentage of hairs with length exceeding 3 mm is found to be linearly related to the linear density of the yarn; that is, there are more hairs in a fine yarn than in coarse yarns of the same type.

Causes of yarn hairiness

a. Raw material

i. *Maturity:* In case of cotton 70% maturity of the fibres is needed; if less it then leads to hairiness. Immature and dead fibres result in hairiness.

ii. *Micronaire (fineness):* Range of micronaire value of fibre for cotton yarn is 4.2 ± 0.2. Micronaire value of 4.2 ± 0.6 results in increased hairiness, which is used in soft, flannel-like fabrics.

iii. *Uniformity ratio:* Normally, uniformity ratio of fibres is 40%-50%. If it is less then hairiness occurs.

b. Process: Inadequate drafting and orientation - spinning triangle.

c. Maintenance: Rubbery surfaces, life of devices or machine parts.

Effects of hairiness

- It causes problems in printing.
- Fine designs are difficult to make.
- Machine parts life is reduced.

Benefits

- Flannel-type fabrics can be produced

Adverse effects of hairiness

1. *Yarn:* Hairiness lowers the yam strength.

2. *Fabric*

Sizing: If yarn is fuzzy, size material will not penetrate to the required amount and a greater amount will accumulate on yarn surface, that is, coating is more but penetration is less.

Shedding problems: In air jet weaving, clear shed is not produced.

Similarly if fabric density is more, two adjacent yarns cause problems due to hairiness.

Wear-out: The machine parts like drop wires, heald wires and reed get worn out frequently.

Knitting: In knitting, the needles get worn out frequently.

Measurement of yarn hairiness

Two parameters are important in the measurement of yam hairiness: the number of hairs in a particular length of yarn; and the length of the hair. The length of the hair can be kept constant for a particular test and the number of hairs may be counted using the Shirley yarn hairiness tester. The instrument consists of a light beam and a photo-receptor opposite to it. The yarn to be tested for hairiness is run between the light and the receptor at a constant speed. As a hair passes between the light and receptor the light beam is momentarily broken and an electronic circuit counts the interruption as one hair. The total number of hairs in a fixed length of yarn is counted for a given time, with the yarn run at a pre-set speed.

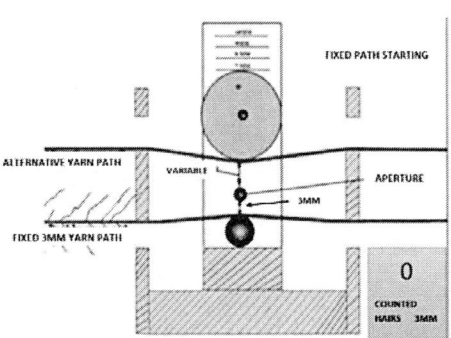

Figure 3.22 Shirley yarn hairiness tester

4.1   Fabric particulars - length, width, crimp, weight, cover factor

Ends per inch and picks per inch

It is a measure of thread density in the fabric. The normal method used to determine thread density is to use a pick glass.

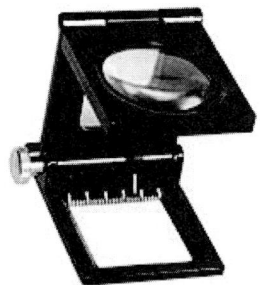

Figure 4.1 Pick glass

EPI and PPI affects the compactness of the fabric. It is also known as thread count or cloth count. Thread counts range from as low as 20 threads per inch as used in tobacco cloth to as high as 350 threads per inch, which is found in typewriter ribbon fabrics. Normally EPI and PPI of a fabric are described as EPI × PPI. Thus, a fabric of 74 × 66 means 74 EPI and 66 PPI.

Sampling

1. Avoid sampling within 50 mm from the selvedge.

2. Avoid sampling within two metres from either end of a piece or roll.

3. While sampling from design fabrics it is convenient to

    a. determine the number of units in a weave repeat from a point paper diagram, and

    b. count

        i. the number of whole repeats

        ii. the remaining units, in the distance across which the threads are to be counted

iii. using the data so obtained, the number of threads per centimetre or inch both in warp way or weft way as required.

Procedure

1. Keep the test sample on a flat table and smoothen it out.

2. Place the counting glass on the fabrics in a direction parallel to warp if weft density is to be determined and parallel to weft if warp density is to be determined.

3. Find the number of warp or weft threads in a specified length as required.

4. Following the procedure prescribed in steps 1 to 4, determine the number of warp and weft threads per centimetre or inch in at least four more places.

5. Calculate the number of warp or weft threads per centimetre or inch by the following formula: $n = N L$

   where

   n = number of warp or weft threads per centimetre (or inch),

   N = observed number of threads in the distance L and

   L = distance in centimetre (or inch) across which the threads are counted.

6. Calculate the mean of all the values and report it as the number of warp or weft threads per centimetre or inch of the fabric.

Crimp

Crimp is defined as the ratio of difference of length of yarn taken from a length of fabric to the length of fabric. Due to interlacing of warp and weft threads, a certain amount of waviness is imparted to the warp and weft yarn in fabric. This waviness is called crimp. Hence, the apparent length of yarn as it exists in the fabric is less than its original or straightened length. A crimp will normally give values ranging from 0.01 to 0.14 (i.e. 1% to 14%).

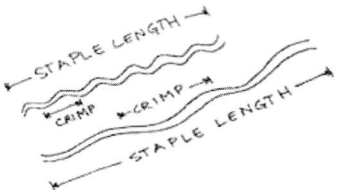

Figure 4.2 Crimp

Crimp is related to many aspects of fabric. It affects the cover, thickness, softness and hand of the fabric. When it is not balanced it also affects the wear behaviour and balance of the fabric because the exposed portions tend to wear at a more rapid rate than the fabric. The crimp balance is affected by the tensions in the fabric during and after weaving. If the weft is kept at low tension while the tension in warp directions is high, then there will be considerable crimp in the weft and very little in the warp.

Measurement of crimp

Percentage of crimp is defined as the mean difference between the straightened thread length and the distance between the ends of the thread in the cloth. From the definition of crimp two values must be known - the cloth length from which the yarns are removed, and the length of the straightened thread. In order to straighten the thread, tension applied must be just sufficient to remove all the kinks without stretching the yarn. In practice it is seldom possible to remove all the crimps before the yarn itself begins to stretch. From those two values we can calculate the crimp percentage with the following formula:

Crimp %, C = (L2 - L1) / L1 × 100

where, L2 = de-crimped yarn length and L1 = crimped yarn length in fabric.

Figure 4.3 Crimp tester

The five groups of threads selected for the test are two-warp-way and three-weft-way groups. The mean crimp percentage is calculated for warp way and weft way. Rectangular strips are carefully marked on the cloth, and each strip cut into the form of a flap. From each strip ten threads are removed. Removal of threads is carried out as follows: the central part of the first thread is separated from the flap fringe by means of a dissecting needle, but the two extreme ends are kept secured. One end is then removed and placed in one grip of the tester, and the other end is removed and placed in the second grip. In this way, the thread is transferred from the cloth to the crimp tester without loss of twist and with minimum handling.

Procedure

- Place the instrument on a horizontal surface and adjust the levelling screw so that it lies flat on the surface.

- The instrument has a tension arm with a rider that moves on the arm. The rider should be kept at the tension value marked on the tension scale. The tension to be applied is calculated using the formula, Tension = Tex / 2. Set the rider against the calculated tension.

- Mark a known length L1 on the fabric for which crimp has to be tested.

- Fix the one end of the yarn, at the mark, on the jaw of the tension arm.

- Fix the other end of the yarn on the jaw provided in the sliding unit.

- Move the sliding unit over the scale, seeing the reference marks through the mirror.

- Stop moving the rider when the reference marks coincide and note the scale reading. This is the extended or de-crimped length L2.

- Calculate the crimp percentage using the formula given above. The procedure may be repeated for all the test specimens. Warp crimp and weft crimp are calculated separately and results are established.

Weight

The weight of a fabric can be expressed in two ways, either as the 'weight per unit area' or the 'weight per unit length'; the former is self-explanatory but the latter requires a little explanation because the weight of a unit length of fabric will obviously be affected by its width. In woven fabric, the weight per unit length is usually referred to as the 'weight per running yard'. It is necessary therefore to know the agreed standard width upon which the weight per running yard is based. Usually this width depends on the loom width. The term that was in use before the arrival of the term 'GSM' (grams per square metre) was 'lb/100 yards'. This expression is used by British Standard. For measuring this there is a template and a quadrant balance. The template area is 1/100 square yards of which each arm is 1/10 yards in length. For measuring GSM, a GSM cutter is used to cut the fabric and weight is measured using a balance. Both of these measurements and methods are equally used for both woven and knitted fabrics. By using GSM it is possible to compare the fabrics in unit area, that is, which is heavier and which is lighter.

Figure 4.4(a) GSM cutter

Figure 4.4(b) Weighing balance

Procedure

Cut five swatches from different places of the fabric. Weigh all test swatches accurately and calculate average weight of swatches. Multiply average weight of swatches by 100 to calculate the GSM of the sample fabric. (Normally, area of round GSM cutter is 1/100 square metres.

In case where a GSM cutter is not available, it is still possible to measure fabric weight per square meter by using following formula.

GSM = Weight of the sample in gram x 10,000 / Area of sample in cm^2

Cover factor

Cover factor is a number that indicates the extent to which the area of a fabric is covered by one set of threads. For any woven fabric, there are two cover factors: a warp cover factor and a weft cover factor. Under the cotton system, the cover factor is the ratio of the number of threads per inch to the square root of the cotton yarn count.

Warp cover factor = EPI / V warp count

Weft cover factor = PPI / V weft count

Cloth cover factor = Warp cover factor + Weft cover factor - (Warp cover factor × Weft cover factor / 28)

4.2   Fabric strength - fabric tensile strength tester, tearing strength tester and hydraulic bursting strength tester

Fabric tensile testing method

Introduction to tensile testing, terminology, CRT, CRL and CRE principles of tensile testing have been explained under yarn testing in Chapter 3. Fabric tensile strength can be tested using universal tensile strength testing equipment with suitable sample mounting jaws. There are different methods of tensile testing of fabric, which are as follows:

1. *Strip test*

    i. Cut strip test

    ii. Ravelled strip test

For ravelled strip test, the specimen is cut wider than the specified testing width, and an approximately even number of yarns are removed from the each side to obtain the required testing width. The ravelled strip test is applicable to woven fabrics, whereas the cut strip test is applicable to non-woven fabrics and dipped or coated fabrics.

The ravelled strip procedure is applicable for the determination of the force required to break a specific width of fabric. The breaking-force information on woven fabrics is particularly useful for comparison of the effective strength of yarns in the fabric with the combined strength of an equal number of the same type of yarns which are not woven. The procedure is not recommended for fabrics having less than 20 yarns across the width of the specimen.

The cut strip procedure is applicable to heavily filled fabrics, woven fabrics that cannot be readily ravelled, felted fabrics and non-woven fabrics. This procedure is not recommended for fabrics which cannot be ravelled as yarns as the edges tend to unravel during testing.

2. *Grab test*

This method is used to determine the effective breaking strength and elongation of woven, non-woven and felted fabrics using the principle of constant rate of extension. This test method is not recommended for glass fabrics, knitted fabrics and other textile fabrics having more than 11% elongation. There are two types of test methods to identify the grab strength. They are

    i. grab test

    ii. modified grab test

These two testing methods are alike, but smaller changes are required in sample preparation.

Sample preparation

Cut strip test

For the fabrics that are difficult to ravel the edges, cut strip method is recommended. Fabric specimen is extended in a direction parallel to the warp or weft, cut to a size of 50 mm x 150 mm or 25 mm x 150 mm exactly. Ensure that the threads of sample run through full length till clamping and accuracy of width. Gauge length should be 75 mm.

Ravelled strip test

A fabric sample is extended in a direction parallel to the warp or weft; make the mark in the fabric using pointed pen to a size of 60 mm x 150 mm or 35 mm x 150 mm. Then fray down the width equally at both sides to get test specimens that are exactly 25-mm wide. Gauge length should be 75 mm.

Grab test

A fabric sample is extended in a direction parallel to the direction of warp or weft. Using an indelible fine marker mark to the size of 100 mm x 150 mm and cut it accurately.

Modified grab test

Mark a cut 37.5 mm from the middle edges of the long dimension. This cut is not necessary to the normal grab test, but mostly modified grab test is preferred for textile fabrics to determine the effective strength.

Figure 4.5 Universal tensile strength tester

Procedure

- Mount the top and bottom jaws in their position properly. Ensure that the jaws are firm and tight. Gauge length should be 75 mm.

- Set the parameter for corresponding test in the instrument.

- Secure the prepared fabric specimen in suitable jaws. Press the test key on control unit.

- After completion of test the result will be shown in the control unit; note this reading.

- After the test results are displayed, bottom jaw returns back to its original position. Now the instrument is ready to test the next specimen.

- Repeat the test for all other warp and weft way specimens. Find the average and the results are expressed separately for warp and weft ways.

Tearing strength

The resistance offered by a textile material when it is subjected to sudden force is generally termed as *tearing strength*.

Figure 4.6 Elmendorf tearing strength tester

Sample preparation

- Mark the specimen with the help of the template supplied with the equipment.

- Cut the warp set with short dimension parallel to warp yarns.

- While cutting test specimens take care to align the yarns running in short direction parallel to template.

- Each specimen of warp set should have different warp yarns, and each specimen of weft set should have different weft yarns.

- Prepare at least five specimens of each type.

Formula

Mean tearing strength (gf) = K × mean value of scale reading / n

where

n = number of test specimens tested together.

k = 16 without any augmenting weight for 1600 gf range.

k = 32 with any augmenting weight for 3200 gf range.

k = 64 with both augmenting weight for 6400 gf range.

Procedure

- Place the tear strength tester on a rigid horizontal table, set the zero point. Next, check the calibrated weights.

- Raise the pendulum sector on the starting position. Secure it by the pendulum sector release mechanism and set the pointer against its stop.

- Clamp a specimen securely in the clamps, so that it is centered with the bottom edge, set against the stop and the upper edge parallel to the top of the clamps; close the clamps by tightening the setting screws.

- Where applicable operate the knife to make the initial slit in lower edge of the specimen. The specimen should lie free with its upper edge directed towards the pendulum sector so as to ensure a shearing action with the pendulum raised and locked against the release lever; fix a dummy test specimen in the grips such that its lower edge is in contact with the fabric, raising it till the knife goes up to maximum extent possible.

- Measure the cut position of the test specimen. If this size is not 20 mm then adjust the blade movement stopper till the cut length is 20 mm.

- Tighten the grips in this position. When waded, the specimen should have its longer edges parallel to top of the grips and the width-wise yarns should be exactly perpendicular to the top edge of the grips. The two grips should hold the specimen under approximately the same force.

- Depress the pendulum sector release mechanism as far as it goes and hold it down until tearing is completed. Catch the pendulum sector by the handle on the return swing, disturbing the position of the pointer.

- Note the position of the needle as indicated by the nearest whole scale division for the capacity used.

- Repeat the operation on the remaining test specimens.

- If the specimen slips through the gaps or if the tear strength deviates by more than 6 mm away from the projection of the original slit, reject the reading and repeat the test with fresh specimen.

Bursting strength

Bursting strength may be defined as the hydrostatic force required to rupture a fabric material. A specimen of the fabric is clamped over an expandable diaphragm which is expanded by the pressure to the point of specimen rupture. The force is applied equally in all directions on the specimen.

Figure 4.7 Bursting strength tester

Sample preparation

Prepare 10 specimens of size 10 cm x 10 cm representing a broad distribution across the width and the length. Ensure that specimens are free from crease or wrinkles. Avoid getting oil, water, grease and so on, on the specimen. The specimen is cut to have a minimum dimension, at least 20% greater than the dimension of the clamp being used. Do not take the test specimen nearer to the edges than one tenth of the width.

Procedure

- The bursting strength tester has two jaws, the fixed lower jaw and the movable upper jaw. The upper jaw can be rotated anticlockwise to open and clockwise to close.
- Place the test specimen between the two jaws and rotate the upper jaw to fit tightly.
- The tester is provided with dual pressure gauge, one reading 0 to 1 kg/ cm^2 and the other 0 to 28 kg/cm^2. Suitable pressure gauge is selected according to the type of the fabric to be tested.
- Press the push button switch and wait until the fabric bursts.
- Note the reading in the pressure gauge when the fabric bursts.
- Repeat the test for other specimens.
- Tabulate the result and find the mean value.

4.3   Fabric abrasion - Martindale abrasion tester

Abrasion is the wearing of any part of the material by rubbing against another surface. Abrasion may be classified as follows,

1. Flat (or) plane abrasion

2. Edge abrasion

3. Flex abrasion

Abrasion resistance

It is the ability of the fabric to withstand rubbing it gets in everyday use. Rubbing may be against itself or another surface. Abrasion test is the standard test method to determine the abrasion resistance of the fabric which is dependent on the fineness of the fabric, the amount of twist of the yarn and the weave structure of the fabric. Yarns that have a firmer and finer twist are generally more resistant to abrasion. The abrasion of the fabric can be tested using Martindale abrasion tester by two methods:

1. Measuring the number of rubs to end-point

2. Calculating the average weight loss

Figure 4.8 Martindale abrasion tester

Sample preparation

Use the standard abrading felt and abrading fabric of ϕ140 mm supplied with the equipment. Cut four numbers standard polyurethane foams of ϕ38 mm with the help of the sample cutter. Prepare 4 specimens of fabric of ϕ38 mm and weigh them before testing.

Procedure

- Remove the top moving plate of the abrasion tester and remove the clamping rings by opening the nut.
- Place the standard abrading felt 140 mm on the abrading platform and cover it with the abrading fabric of ϕ140 mm.
- Place the 9 kp load in the centre of the abrading platform and clamp it with clamping ring nuts.
- Check for tucks and ridges and place the top moving plate carefully and correctly.
- Prepare the specimen-holding units as follows:
  a. Put the bottom unit in the specimen holder opener.
  b. Place the testing fabric of ϕ38 mm into the bottom unit in such a way that the testing face of the fabric should face downside.
  c. Put the polyurethane foam of ϕ38 mm on the fabric.
  d. Place inside unit of specimen holder in it.
  e. Tighten the top cover by exerting slight pressure on it.
- Press the complete specimen holder on the center of the abrading platform.
- Place the specimen holding brush on the top moving plate and tighten it with the nut. Add the 9 kp weight if the fabric to be tested is for apparel use.
- Set the number of cycles as required (e.g., 50, 60 etc.).
- Start the machine and wait until the set number of cycle is complete.
- Remove the sample from the holder and weigh it.
- The abrasion resistance of the fabric sample is calculated using the following formula:

Abrasion resistance % = (Abraded weight / Original weight) × 100

Also weight loss can be calculated as follows:

Weight loss % =

[(Original weight - Abraded weight) / Original weight] × 100

4.4   Fabric pilling - ICI pill box tester

Pilling

The entanglement of fibres, generally termed as *pills,* stands protruding on the fabric surface during wear and tear. A test to assess the property of pills that form and or can be retained on the surface of a fabric when it is subjected to specified conditions is called pilling test.

Figure 4.9 ICI pill box tester

Sample preparation

- Place the fabric facing downwards on a plain surface, and on it place the template with its longer edges along the weft direction. Draw lines using a pencil along the edges and in the slits of the specimen template.

- Fold the specimen to face inwards until the longer edges touch each other and sew exactly along the inner pencil lines. Cut from the sewn sample and specimen along the length, each 125 mm long. Then turn the specimen inside out so that the face side of the fabric is outside.

Mounting the test specimen

Take a pilling tube and the specimen holder. Place the pilling tube in the specimen holder, from its open side, by pushing it inside. When the pilling tube goes inside about half its size, push the specimen over the metal cylinder from the other side of metal cylinder and place it on pilling tube. Withdraw the cylinder and the pilling tube from the jig and allow it to recover to its original circular configuration with the specimen wrapped around it under even tension. Use the PVC tape to wrap around the ends of the fabric to secure the ends.

Procedure

- Clean the boxes thoroughly. Place four mounted test specimens in each box and close the boxes.
- Set the machine for 1800 revolution with help of keypad buttons. Press the start button.
- The machine would stop automatically after 1800 revolutions.
- Take out the specimens carefully and compare them with the photographic rating standards.

4.5   Fabric drape - measurement by drape meter

Fabric drape is one of the important properties of appearance of a garment during wear. It is a complex property including bending and shearing deformation. Drape is important for the selection and development of textile material for apparel industry. Fabric drapability may be described as a degree of the deformation of fabric to orient itself into folds, when the fabric is partially supported by other objects.

Drape coefficient:. The drape of the fabric is measured in terms of drape coefficient F (%) using the formula.

Drape coefficient, F (%) = [(D - a) / (A - a)] × 100

where

a = area of supporting stand of 12.5 cm diameter $(122.8 \text{ cm})^2$

A= area of test specimen of 25-cm diameter $(4.91 \text{ cm})^2$

D = area of drape pattern is calculated as shown below:

Area of drape pattern (D) = K × w/W

where

K= correct factor of divergence of light rays (0.91)

w= mass of the drape pattern

W= mass of ammonia process paper in grams per square cm

Figure 4.10 Drape meter

Sample preparation

Select the location for cutting the test specimen such that the areas within 5 cm of the selvedge or those having wrinkles or sharp folds are not included in the specimen cut out. Place the fabric under the test on a flat table such that the locations from which the test specimens are to be taken are free from any crease or wrinkles. Place the specimen template over selected location and trace the outline and mark the centre cut in the specimen using a pair of sharp scissors. Fold the test specimen in four quarters about centre; strip off the corner to leave a hole of 5-mm diameter.

Procedure

- Measure the dimensions of the acrylic sheet in the Drapemeter and cut an equal size of ammonia paper and weigh it and note down the weight. Determine the mass per unit area of the paper used by cutting a known area of the original paper and weighing.

- Raise the hinged, acrylic sheet plate and place the ammonia process paper, face up, over the resilient black sheet below the acrylic sheet.

- Lower the acrylic sheet so that it sits firmly on the paper. Ensure the paper lies flat and is free of any fold (or) crease.

- Remove the specimen-loading assembly from the specimen support. Place the specimen between the plates of this assembly with the bolt passing through the central hole of the specimen.

- Tighten the thumbnail. Hold the thumbnail of the assembly and briskly move the holder with the specimen up and down 10 times, each time resting on the table for a moment. This is to allow the fabric to orient freely and drape into natural fabric.

- Lower the test specimen holder assembly with the test specimen gently over the specimen support such that the head of the bolt sits in the hole at the centre of the test specimen support.

- Switch on the halogen lamp tube. Note the time at which the lamp has been switched on; allow the paper to get exposed for a period of 10 to 12 minutes.

- Open the door of the developing chamber below the exposing chamber.

- Remove the bowl placed inside, if the level of ammonium hydroxide in the bowl is less than half and more liquor in the bowl does not give of a strong smell of ammonia; drain out the old liquor and add fresh liquor to fill three-fourth of the bowl.

- Keep the bowl with liquor ammonia in the developing chamber.

- Set the duration of exposure, for 20 minutes and turn on the lamp.

- Remove the exposed paper from the exposing chamber and place it with the exposed face downwards over the welded wire melt platform inside the developing chamber. Allow it to remain in this state for 10 minutes.

- After 10 minutes, remove the developed paper from this chamber.

- The outline of the exposed area on the paper will be slightly fuzzy or diffused because of divergence of light rays.

- Draw freshly a smooth curve through the centre of the diffused areas, leaving out fuzzy portions or indistinct boundaries.

- Condition the paper to moisture equilibrium at standard atmosphere and cut out the drape pattern with a pair of scissor and determine its mass in gram corrected to two decimal places.

- Reverse the specimen and obtain the pattern with the outer surface upwards.

- Test at least four specimens making a total of eight measurements.

- Calculate the drape coefficient separately for face and back of the fabric.

4.6   Fabric stiffness - Shirley stiffness tester

Stiffness is the resistance offered by material to a force which tends to bend it. It is related to the handle and drape of fabric. The measure of the draping quality of the fabric is called bending length. It is calculated using the following formula.

Bending length, C = [L (cos$\frac{\theta}{2}$)$^{1/3}$] / 8 tan θ cm

Where $\theta = 41^{1}/_{2}{}^{\circ}$, the standard angle of deflection.

L = length of fabric at which the strip coincides.

The measure of stiffness associated with handle is flexural rigidity. It is calculated using the following formula:

Flexural Rigidity, G = W x C^3 gm cm

where W = cloth weight in grams/sq cm

C = bending length in cm

Also, overall flexural rigidity (G) is calculated using the formula,

G = (G$_1$ + G$_2$)$^{1/2}$ gm-cm

where G$_1$ = flexural rigidity in the warp way

G$_2$ = flexural rigidity in the weft way

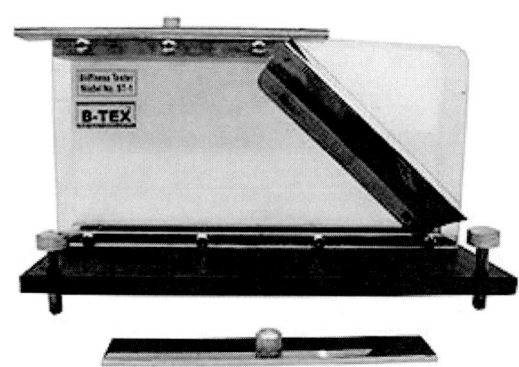

Figure 4.11 Stiffness tester

Sample preparation

Prepare rectangular warp-way and weft-way test specimens of 200 mm × 25 mm size with the help of the template from different portions of the sample under test. The lengthwise direction of specimens should be parallel to the warp or weft direction for which the stiffness is to be determined. Specimens cut in each direction shall be scattered on as far as possible, so that no two warp-way specimens contain the same set of warp yarn and no two weft-way specimens contain the same set of weft yarns. Avoid selvedge and portions with creased or folded places. The specimen should be handled as little as possible.

Procedure

- Place the instrument on a table or bench so that platform is horizontal and inclined reference lines are at the eye level of the operator.

- Adjust the platform with the help of a spirit level so that it is horizontal.

- Place one of the specimens on the platform with the scale on the top of the specimen and zero of scale coinciding with the leading edge of the specimen.

- Holding the scale in the horizontal plane, start pushing the specimen and scale slowly and steadily.

- When the leading edge projects beyond the edge of the platform the fabric tends to drop over the edge and bend under its own weight.

- Push the fabric and the scale until the tip of the fabric cuts both the index lines when viewed in the mirror and stop pushing.

- Note down the length of the overhanging portion from the scale to the nearest millimetre.

- Take four readings from each specimen with face and back side up, first at one end and then at the other.

- Similarly, test at least five specimens for each warp way and weft way.

- Calculate the flexural rigidity along the warp, weft and the overall flexural rigidity.

4.7   Fabric crease resistance and crease recovery – measurement of crease recovery

An unintentional fold in a fabric that may be introduced at some stage in processing is called crease. The resistance offered by a textile material for creasing during use is called crease resistance. The crease resistance is measured quantitatively in terms of crease recovery angle.

Figure 4.12 Crease recovery tester

Sample preparation Cut warp-way and weft-way test specimen measuring 15 mm × 40 mm in size with the help of a sharp pair of scissors with their longer side parallel to warp and weft threads, respectively. The specimens shall be staggered in such a way that no two warp-way specimens contain the same set of warp yarn and no two weft-way specimens contain the same set of weft yarns. There shouldn't be any wrinkles, bends or other deformations in the samples chosen as well as in the area within 50 mm from the selvedge. Prepare five such specimens each for warp and weft way.

Procedure

- The specimen is carefully creased by folding it in half; then it is placed in the gap in the creasing plate and a creasing load weighing 1 kg is kept over the specimen for a minute.

- After 1 minute, the load is removed and the specimen is transferred to the measurement device where one end of the specimen is held in a fabric clamp and the other end is allowed to fall free under its own weight and allowed to recover from the crease.

- As it recovers the dial of the instrument in rotated such that the free edge of the specimen is in line with the knife edge.

- At the end of the time period allowed for recovery, which is usually a minute, the recovery angle in degrees is read on the scale.

- This angle gives the measure of the crease recovery angle.

- Tabulate results separately for warp-way and weft-way specimens.

- Find the mean of the results to the nearest degree.

- The load time or creasing and recovery time may be altered to suit particular cases.

4.8   Fabric permeability - Shirley air permeability tester, fabric permeability to water, Bundesmann tester

Permeability may be defined as the rate at which gas or liquid passes through a porous medium. Textile fabrics are permeable substances. The fabric needs air, water and vapour permeability so that a person feels comfortable wearing it.

Air permeability is defined as the rate of air flow passing perpendicular through a known area under a prescribed air pressure differential between the two surfaces of a material. It is a measure of how well a fabric allows the passage of air through it. Apart from apparel comfort, it is also important for a number of fabric end-uses, for example, industrial filters, tents, sail-cloths, parachutes, air bags and so on. The level of air permeability varies depending on the following:

- Type of yarn
- Fabric structure
- Fibre parameters

Important fabric properties for maintaining thermal comfort include the following:

- Air permeability
- Water or moisture (vapour) permeability/transportation
- Heat transmission

Air permeability affects the comfort aspect of a garment in terms of air passage through the fabric. High air permeability per unit area of a fabric gives lower protection against winds, especially for outer-wear garments, whereas low air permeability causes heavy body perspiration.

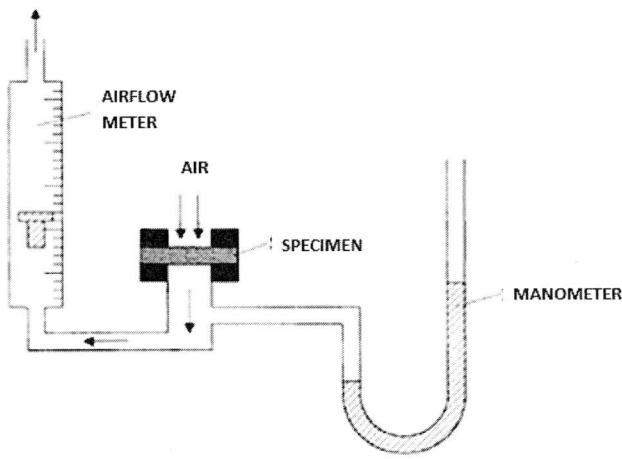

Figure 4.13 Shirley air permeability tester

In Shirley air permeability tester the sample is clamped between two rubber gaskets, and a guard ring surrounding the test specimen ensures that all the measured airflow passes through the specimen with no leakage. The test area is a circle of 5.07 cm2. Airflow is measured when a pressure differential of 20 mm H20 (0-2 kPa) is applied. Ten measurements of airflow are made on each sample.

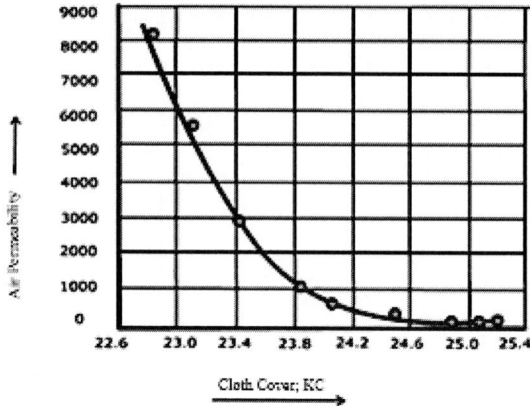

Figure 4.14 Relationship between air permeability and cloth cover

Procedure

- Handle the test specimens carefully to avoid altering the natural state of the material.

- Place each specimen onto the test head of the test instrument and perform the test as specified in the manufacturer's operating instructions.

- Place coated test specimens with the coated side down (towards the low-pressure side) to minimize edge leakage.

- Use a water pressure differential of 125 Pa (12.7 mm or 0.5 in. of water).

- Read and record the individual test results in SI units as $cm^3/s/cm^2$ and in inch-pound units as $ft^3/min/ft^2$ rounded to three significant digits.

- For special applications, the total edge leakage underneath and through the test specimen may be measured in a separate test, with the test specimen covered by an airtight cover, and subtracted from the original test result to obtain the effective air permeability.

- Remove the tested specimen and continue testing until all the specimens have been tested for each laboratory sampling unit.

- The number of tests may go up to 10 but the minimum required number of tests is 4.

Water permeability

Perspiration is an important mechanism which the body uses to lose heat as its temperature starts to rise. Perspiration is in two forms:

(a) Vapour form - passes through the air gaps between yarns in fabric

(b) Liquid form - occurs at higher sweating rates and it wets the clothing which is in contact with the skin.

Water vapour permeability test

The specimen under test is sealed over the open mouth of a dish containing water and placed in standard testing atmosphere. Total weight at the start of test is taken (W_0). After specified time, the weight of setup is taken as W_t. The rate of water vapour transmission is calculated from the difference between W_0 and W_t [water vapour permeability (W_{VP})].

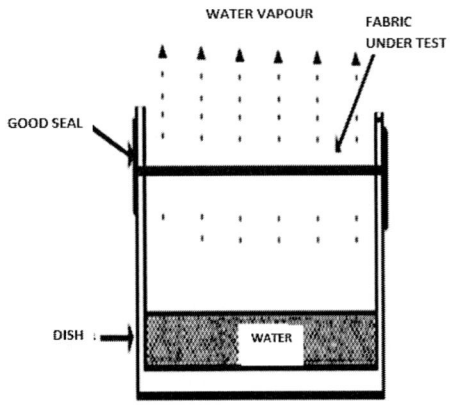

Figure 4.15 Water vapour permeability tester

Water permeability - Bundesmann tester

The Bundesmann water repellence tester is used to determine resistance to wetting and resistance to penetration of water-repellent fabrics which are permeable to air. In this test, four test specimens taken from the fabric under test are simultaneously exposed to a simulated heavy shower of controlled intensity while the lower surface of each specimen is subjected to a rubbing action. The increase in mass of specimen during exposure to shower is determined and any water passing through the fabric is collected and its volume is determined.

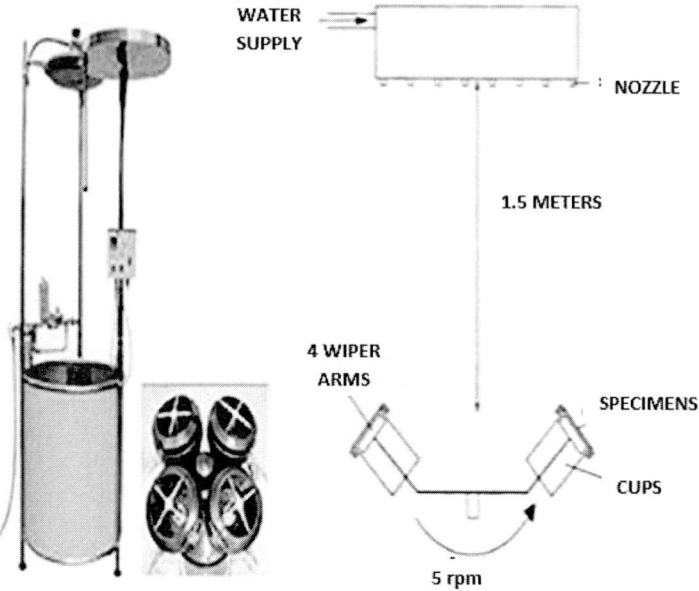

WATER SUPPLY

NOZZLE

1.5 METERS

4 WIPER ARMS

SPECIMENS

CUPS

5 rpm

Figure 4.16 Bundesmann water repellence tester

Two fabric parameters are determined from the test:

1. Penetration of water through the fabric: the water collected in the cups is measured to the nearest ml.

2. Absorption of water by the fabric: in order to do this the specimen is weighed before the test and then after the shower. To remove excess water the fabric is shaken 10 times using a mechanical shaker and then weighed in an airtight container.

Absorption % = (Mass of water absorbed / original mass) × 100

In each case the mean of four values is calculated.

4.9   Colour fastness to rocking, perspiration, sunlight, laundering, dry cleaning and hot pressing

Colour of a textile material refers to the dye that is used to impart certain colour or any pigment that is used to print on the fabric. These dyes or pigments tend to fade over time and also because of repeated washing. So it is important to test the fastness of any dyed or printed textile in order to assess the quality of the dye or pigment being used. Colour fastness is the property of dyed textile material to resist any named agency, for example, washing or laundering, light, crocking or rubbing, perspiration and so on.

Rubbing or crocking fastness is the transfer of dyed textile material's colour from its surface to an adjacent area by the same fabric or other surface (normally bleached cotton fabric) principally by rubbing. It is tested using a crockmeter. It may be either hand-driven or motorized. This test is designed to determine the degree of colour which may be transferred from the surface of a coloured fabric to a specific test cloth for rubbing (which could be dry and wet). Fastness to rubbing is a main test which is always required for every coloured fabric, be it printed or dyed.

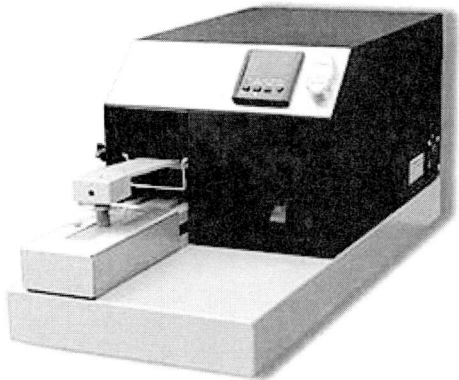

Figure 4.17 Crockmeter

Sample preparation

Prepare two pieces of fabric measuring 5 × 13 cm, one for dry rubbing and the other for wet rubbing. The fabric is cut along the warp so that the longer side is parallel to the direction of warp. In case where the test sample contains more than one colour, multiple samples covering all the colours are prepared and crocking test is performed on each colour.

Procedure

- Prepare the test specimen and fasten it to the base board of the testing device by means of clamps. Care should be taken to ensure that there are no folds or creases on the surface.

- When testing multicoloured textiles care should be taken to position the specimen in such a way that all colours of the design are rubbed in the test. Alternatively, if the areas of colour are sufficiently large, more test specimens may be taken to assess individual colours separately.

- Initially the crock meter is set at 10 strokes, the stroke length is kept as 100 mm and the downward force of the fingers is set as 9 N for normal fabric and the diameter of the finger is 16 mm.

- Dry rubbing: Mount the dry cotton rubbing cloth (5 cm × 5 cm) on the finger and fasten with a clamp. Start the crock meter and when the test is complete the fabric sample and the rubbing cloth are taken and kept aside for assessment.

- Wet rubbing: Wet the cotton rubbing cloth in distilled water, squeeze out excess water by using blotting paper so that it absorbs about 65+5% of water. Now mount this wet rubbing cloth on the finger and repeat the testing. After completing the test, dry the wet rubbing cloth in room temperature and then assess the rubbing cloth.

- Use the grey scale for assessing staining on the rubbing cloth and assign grades accordingly.

Grey scales for assessing colour change and staining

There are two types of grey scale - one for assessing 'colour change' and another for 'staining'. The colour-change scale consists of nine pairs of grey- coloured chips, in grades ranging from 1 to 5. The staining scale consists of nine pairs of grey and white-coloured chips, in grades ranging from 1 to 5. They are manufactured strictly in accordance with ISO 105-A02 and AO3.

Grey scale for assessing staining

A scale that consist of one pair of white and nine pairs of non-glossy grey and white-coloured strip and in which a fastness rating of 5 (indicating no staining) at one end of the scale is represented by the pair of white chips, and the fastness rating of 1 (indicating severe staining) at the other end is represented by a pair of chips with one complete white and one complete grey, which shows a huge difference. The fastness rating goes step-wise from

Note 5 = no visual change (best rating) to Note 1 = a large visual change (worst rating).

The grey scale has nine possible values: 5, 4-5, 4, 3-4, 3, 2-3, 2, 1-2, 1.

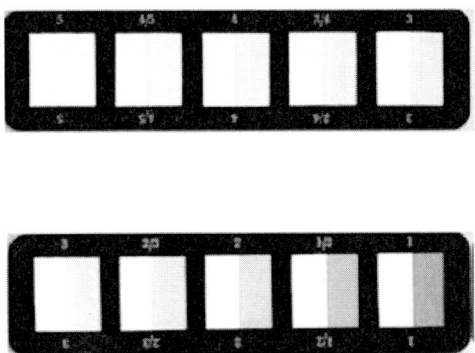

Figure 4.18 Grey Scale for assessing staining

Assessment of specimens

Assessment has to be done independently by two persons and the average is taken as result. The assessment has to be done with defined grey scale, grey frame in same grey colour as background, white paper as background or according to the test method one or more original multi-fibre fabric to ensure that no transparent effect comes out of the background and under defined light D65 from a defined angle of 45° using the colour matching cabinet.

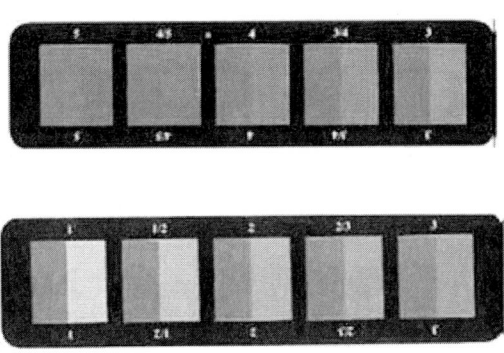

Figure 4.19 Grey Scale for assessing colour change

Colour fastness to washing

The colour fastness to washing is tested by assessing the colour loss and staining on adjacent fabric resulting from desorption and or abrasion action in one single test that is closely approximate to one commercial or domestic laundry. It is tested using the equipment called Laundrometer or wash wheel.

Figure 4.20 Laundrometer

Multi-fibre fabric

Multi-fibre fabric is used to assess colour transfer in colour fastness testing. Fabric to be assessed will be 100 mm (4 inch) wide, with warp stripes of 6 fibres as shown below. Fabrics are usually available in 10 metre rolls or as cold cut pieces with pinked edges. Filling stripes' measurements are as follows: 10 cm (4 inches), repeat 45 inches width × 1 yard. There are different types of multi-fibre fabric that can be used based on the testing standard to be adopted. The multi-fibre fabric used for ISO standard test is as follows:

- Acetate
- Bleached cotton
- Spun nylon 6.6
- Spun polyester (Dacron 54)
- Spun acrylic (Dralon)
- Worsted wool

Sample preparation

A 100 mm × 40 mm test specimen is taken and attached to a 100 mm x 40 mm multi-fibre adjacent fabric by sewing along one of the shorter edges, with the multi-fibre placed adjacent to the face-side of the specimen.

Detergent

ECE reference detergent without optical brightener may be used. A minimum volume of 1 litre of detergent solution shall be prepared because of possible lack of homogeneity of the detergent powder.

Procedure

- Prepare the wash liquor by dissolving 4 gm of detergent per litre of water.

- Take a stainless steel jar from the laundrometer and add 150 ml of wash liquor and 10 numbers of steel balls and place the test specimen and close it tightly.

- Fill the laundrometer with adequate water. Set the temperature to 40°C and wash time to 30 minutes.

- Fit the jar in the stainless steel rotor assembly of the laundrometer and start the laundrometer.

- The assembly rotor rotates at the speed of 40 ± 2 rpm. When the test is over the laundrometer stops automatically with a beep. Remove the jar from the laundrometer.

- Take the composite specimen out of the jar and rinse twice for 1 minute in 100 ml of water using fresh water every time (i.e., do not reuse the water used for the first rinse).

- Extract excess water from the composite specimen with the help of absorbent paper or cloth.

- Dry the specimen in air at temperature not exceeding 60°C with the parts in contact only at the line of stitching.

- Assess the change in specimen colour and the staining of the adjacent fabric using the appropriate greyscale.

Colour fastness to perspiration

Perspiration from the human body is a complex chemical containing large quantities of salts. Depending on the human metabolism, it can be either acidic or alkaline. The tests for colour fastness to perspiration are based on the solution prepared by simulating the acid and alkaline perspiration.

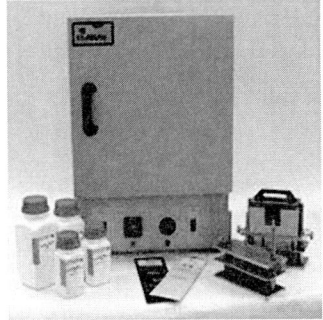

Figure 4.21 Perspirometer and oven

Sample preparation

Cut two specimens (100 × 40 mm) along the length or width of the fabric. For multicoloured fabric or fabric with prints or stripes or checks, all the colours of design should be in contact with all the components of multi-fibre. Where it is not possible for all colours to be in contact with all the components, more than one test specimen may need to be prepared. For fabrics with different shades and fibre compositions on the face and back both sides should be tested. Avoid placing sample within 50 mm of the selvedge.

Test solution preparation

The simulated acidic and alkaline perspiration solutions are prepared using certain chemicals as shown in Table 4.1.

Table 4.1 Simulated perspiration solutions preparation

Chemical	Alkaline Solution	Acid Solution
1 - Histidine monohydro chloride monohydrate	0.5 g	0.5 g
Sodium Chloride	5.0 g	5.0 g
Disodium hydrogen orthophosphate dihydrate	2.5 g	2.2 g
Volume	1 litre	1 litre
pH	8.0	5.5
To obtain the correct pH, 0.1 N Sodium Hydroxide solution is added to the test solution.		
Test Condition		
Temperature	37±2°C	37±2°C
Time	4 hours	4 hours

Procedure

- Align the test specimen and multi-fibre fabric and sew together along the short edge to form a composite specimen.
- Weigh each composite specimen. Calculate the amount of test solution according to a liquor ratio of 1:50 per specimen.
- Immerse one composite specimen in alkaline solution and the other in acid solution for 30 minutes at room temperature to ensure that specimens are thoroughly wet.
- In the mean time, turn on the heating oven and set the temperature at 37°C.
- Place each specimen onto an acrylic plate and draw across with a stirring rod on top of the specimen to fully enclose it. Drain off any

excess solution. Place some more acrylic plates below and above the test specimen.

- Bring the upper pressure plate of the perspirometer down to the top of the acrylic plates and place the loading weight onto the pressure plate, and then tighten the side screws. This is equal to a pressure of 12.5 kpa (5 kg) in between being applied on the specimens.

- Use two different perspirometers, one for acidic perspiration test and another for alkaline perspiration test.

- Place the perspirometers into the oven, maintaining the temperature at $37\pm2°C$ for 4 hours.

- On completion of the test, remove the perspirometers from the oven and take the specimens out of the plates.

- Open out the specimens so that they are only in contact at the line of stitching.

- Dry the specimen in an incubator or by drying on a screen at a temperature not exceeding 60°C.

- After drying, assess all the specimens using the grey scale for assessing colour change and assess all the components on the adjacent multi-fibre fabric using the grey scale for assessing staining.

Preparation for evaluation

Trim off ravelled yarns and lightly brush off any loose fibre and yarn on the fabric surfaces. For pile fabric specimens, brush the pile in the required direction to restore them as close as possible to the same pile of the original control sample. Specimens should be smoothed and flattened if necessary.

Colour-change evaluation

Compare sample to control fabric. Colour change may also be quantitatively measured using spectrophotometer.

Staining evaluation

Compare sample to control fabric. State the basic conditions of the test (i.e.. sample size, temperature etc.). Report the grade number determined for colour change and the staining grade numbers. State the scales that were used. Report the specific fibres used in the multi-fibre test fabric.

Colour fastness to light

This refers to the ability of the fabric to withstand the sun light. The resistance to sunlight depends on the following:

- The intensity of light
- Inherent properties of the fabric
- Season
- Altitude
- Distance from the equator

Measurement methods

1. Direct method: Sunlight method

   Samples are placed in a glass cabinet and exposed to sunlight from 9 am to 3 pm.

2. Laboratory instrument method

   - Fade-o-meter
   - Weather-o-meter

These machines have either a carbon-arc lamp or xenon-arc lamp as the light source.

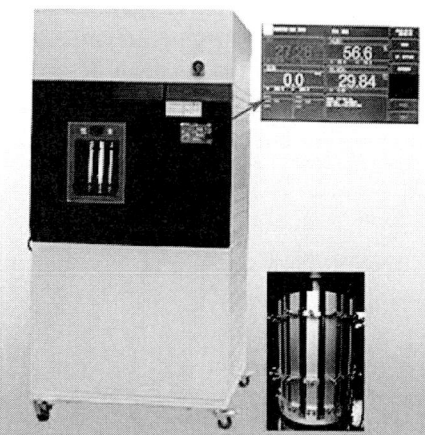

Figure 4.22 Light fastness tester

Blue wool standards

This test measures the resistance to fading of dyed textile when exposed to day light. The test sample is exposed to light for a certain time which is about

24 hours to 72 hours or by customer/buyer demand and compare the change with original unexposed sample. The changes are assessed by grey scale.

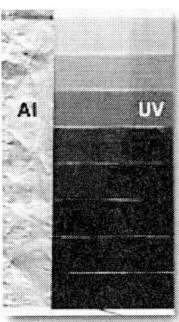

Figure 4.23 Blue wool standards

Procedure

For testing light fastness, samples are exposed to an intense artificial light generated by a Xenon arc lamp. The light passes through a series of filters to ensure that its spectrum (wavelength) closely matches the spectrum of natural daylight coming through the glass. The humidity and temperature of the test atmosphere are controlled. A small sample of material, together with a series of eight blue wool standard fabrics (reference 1–8), are simultaneously exposed to high-intensity light. Two exposure times are used to determine whether fading progresses steadily or initially at a different rate from the longer-term exposure.

The test involves assessing samples once during the test and again at the end. The assessment is completed by comparing the degree of fading shown on the sample with the fading shown by the blue wool standard fabrics. Each sample is assigned a light fastness rating based on the number of the blue wool standard which shows the equivalent degree of colour change.

Light fastness grade

Table 4.2 The sixteen parameters describing fabric mechanical and surface properties

PARAMETER	SYMBOL	DEFINITION	UNIT
Tensile	ET	elongation	%
	LT	linearity	none
	WT	tensile energy	gf.cm/cm²
	RT	resilience	%

Shearing	G	shear rigidity	gf/cm.deg.
	2HG	hysteresis(0.5˚)	gf/cm
	2HG5	hysteresis(5.0˚)	gf/cm
Compression	LC	linearity	none
	WC	compression energy	gf.cm/cm²
	RC	resilience	%
Surface	MIU	Co ef. of friction	none
	MMD	Mean dev. of miu	none
	SMD	Geom . roughness	micron
Bending	B	Bending rigidity	gf/cm²/cm
	2HB	Hysteresis	gf.cm/cm
Fabric construction	W	Fabric weight/ unit area	mg/cm²
	TO	Fabric thickness	mm

Colour fastness to dry cleaning

Dirt or contamination in textiles is often removed by either aqueous cleaning using detergents or chemical or dry cleaning using organic solvents. Dry cleaning is normally used in the case of textiles that cannot be washed for any reason or that cannot be washed by aqueous means, for example, wool and silk garments. Due to dry cleaning, dyed or printed textiles may change their colour to some extent (fade), and another problem is that they discolour the solvent used in dry cleaning.

Procedure

Specimens of textiles in contact with a cotton fabric together with non-corrodible steel discs are agitated in perchloroethylene and squeezed or centrifuged and dried in hot air. The change in specimen colour is assessed with the grey scale for assessing change in colour. Upon conclusion of the test, the staining of the solvent is assessed by comparing the filtered solvent with unused solvent by means of the grey scale for assessing staining.

Colour fastness to hot pressing

Pressing fastness test of dyed and printed textile products is performed to determine the resistance of textiles to ironing and to processing on hot cylinders. Different tests are performed at three different conditions, that is, when the textile is dry, wet and damp.

Procedure

1. *Dry pressing*: The dry specimen is pressed with a heating device of a specified temperature, time and weight. Place the specimen on the piece of dry cotton cloth on a smooth horizontal surface. Place the iron on the specimen and leave it for 15 seconds.

2. *Wet pressing:* The wet specimen is covered with a wet, un-dyed cotton cloth and pressed with a heating device at specified temperature and weight. Soak the specimen and cotton cloth in distilled water and squeeze or extract them to contain their own weight of water. Place the wet specimen on a piece of dry cotton cloth, place the wet cotton cloth on the specimen and press by moving the iron to and fro over the wet cloth (without additional pressure) for 15 seconds.

3. *Damp pressing:* The dry specimen is covered with a wet, undyed cotton cloth and pressed with a heating device of a specified temperature, time and weight. The adjacent fabric in dimensions of 100 mm × 40 mm is immersed into the water and squeezed so as to get 100% pick-up. Dry specimen is placed onto the un-dyed cotton fabric on the filler and the adjacent wet fabric is placed on the specimens. The top layer of the device is pulled down, and dry specimen is pressed for 15 seconds at the determined temperature. The change in colour of the specimen is evaluated in two different ways using grey scale:

   a) Soon after the end of the test

   b) After conditioning for 4 hours under standard atmospheric condition.

The staining on un-dyed fabrics is evaluated using grey scale.

Apparatus and reagent

Hand iron with proper weight to give a pressure of approximately 30 g/sq.cm at temperatures indicated as follows for different fabrics:

- Cotton and linen   190–200°C
- Wool, silk and viscose   140–160°C
- Cellulose acetate and polyamide   115–120°C

Five pieces white bleached cotton cloth weighing approximately 125 GSM, having a size not less than 14×4 cm are required. Gray scale for assessing change in colour and staining (ISO scale). The size of the specimen is 10 × 4 cm.

4.10   Colour matching - colour-matching cabinets and computer colour-matching system

Colour-matching cabinet

The colour-matching cabinet is an instrument designed for visual evaluation of shade and colour difference under standardized and controlled light conditions. It is used for visual inspection of any coloured material like textile, paper, paint, ink, plastic and so on. Standard light sources are used for accurate inspection and evaluation. It is made of teakwood, ply board and so on, with strict adherence to international standards. The interior paint should match with Mussel N7 neutral grey paint and exterior powder coating with metallic finish. It has an inclined viewing stand of 45° angle. There are six different light sources to evaluate the colour variations of the specimen.

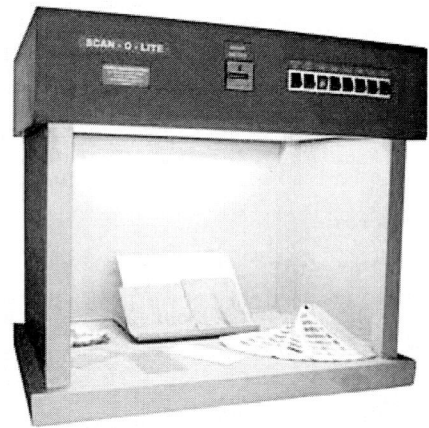

Figure 4.24 Colour-matching cabinet

Light Sources of colour-matching cabinet

Artificial daylight fluorescent lamps (D-65)	: 1 no. (18 W)
Tungsten filament lamps	: 4 nos. (40 W)
Cool white fluorescent lamp (CWF)	: 1 no. (18 W)
Triphosphor fluorescent lamp (TL-84)	: 1 no. (18 W)
Ultra-violet black lamp	: 1 no. (8 W)
Warm white	: 1 no. (18 W)

The equipment offers a wide-viewing area equipped with multiple light sources or lamps to detect the phenomenon of metamerism, where samples appear to match under one light source but distinctly different under another. In order to reach the final approval the fabric has to go through various light sources, which include artificial daylight, cool white light, tungsten filament light, ultra violet black light and triphosphor fluorescent light.

Procedure

- The sample to be evaluated and the original sample are to be kept in the viewing cavity.
- After switching on the D-65 light, closely check the original specimen and the specimen to be evaluated.
- When the match is satisfactory, the specimen is to be evaluated under CWF light and TFL lamps.
- Now the evaluation is carried out under UV-B light; it is necessary to ascertain whether both samples react equally to ultraviolet radiation.
- Then the final evaluation is carried out under TL-84 triphosphor fluorescent lamp tube light to ascertain whether the colours are acceptable.
- All these help in judging the colour differences accurately, by comparing the samples in different conditions. When the match of both samples holds well under all the lights, it means that the dyeing, printing and the colouring of the fabric or yarn are as per the requirement.
- Apart from colour matching, assessment of colour change and staining after colour fastness testing

Computer colour-matching system (CCMS)

CCMS is the instrumental colour formulation based on recipe calculation made by using the spectrophotometric properties of dyestuff and fibres. Generally buyer gives a fabric sample swatch or Pantone number of a specific shade to the manufacturer. Then the manufacturer gives the fabric sample to lab dip development department to match the shade of the fabric. After getting the sample they analyse the colour of the sample manually. It is a laborious, time-consuming and critical task and needs skills and expertise of the personnel developing the lab dip. On the other hand, to save time and money, they can use CCMS.

The three basis components in CCMS

- Colour measurement instrument (spectrophotometers)

- Reflectance (R%) from a mixture of dyes or pigments applied in a specific way
- Optical model of colour vision to closeness of the colour matching (CIE L*a*b).

Functions of CCMS

- Colour match prediction
- Colour difference calculation
- Determination of metamerism
- Pass/fail option
- Colour fastness rating
- Cost comparison
- Strength evaluation of dyes
- Whiteness indices
- Reflectance curve and K/S curve
- Production of shade library
- Colour strength measurement

Working procedure of CCMS

The manual process mentioned above is made easy and faster to execute with the use of CCMS. First, the given sample or pantone shade card is measured using a spectrophotometer. Then it is analysed for the shade and depth of colour. At the same time, it determines the colour combination with which the required colour can be obtained by dyeing the fabric. It generates some dyeing recipes and gives a range of options for the user. The amount of chemicals needed for dyeing is determined. The sample is dyed and measured using the spectrophotometer. Later, it is compared with the original sample given by the buyer. If it matches, then colour-matching is considered a success, and if not, a failure.

Advantages of CCMS

CCMS provides a lot of advantages to the textile Industry.

- Customers get the exact shade they require given the degree of metamerism possible to achieve by CCMS.
- Customers can choose from a number of colour formulations, ranging 10 to 20 formulations that will match a colour. By taking costing, availability of dyes and auxiliaries into account, one can choose a best swatch out of these options.

- It is 3 to 300 times faster than manual colour matching.
- Limited range of stock colour needed.

Whiteness index

Whiteness is a single-number index referencing the relative degree of whiteness (of near-white materials under specific lighting conditions). The index has been devised such that most people agree that the higher the whiteness, the whiter the material. CIE whiteness, developed by the France-based International Commission on Illumination (also called CIE), is the most commonly used whiteness index. This index normally refers to measurements made under D65 illumination, which is a standard representation of outdoor daylight. For a perfectly reflecting nonfluorescent white material, the CIE whiteness would be 100.

Whiteness is associated with a region in the colour space within which objects are recognized as white. Degree of whiteness is in principle measured by the degree of departure of the object from a perfect white. A wide variety of indices are available for those industries that need to evaluate the whiteness of their products, for example, textile, paper and paint industries. It is a measure which correlates the visual ratings of whiteness for certain white and near-white surfaces. The two most commonly used methods for computing a whiteness index are: CIE Whiteness Index and Ganz-Griesser Whiteness Index.

CIE Whiteness Index

To make the white weighting more informative, the CIE recommended in 1981 a formula that is today known as 'CIE Whiteness'. These indices have been specified for the D65 and illuminant C in combination with either 2° or 10° observer function. However, the equation is commonly used with other illuminants; therefore, the value shown will depend on the primary illuminant being chosen.

Granz-Griesser Whiteness Index

Tint-CIE whiteness is a quantifying parameter based on the white evaluation by Ganz and Griesser. It is a simplified version of the Ganz-Griesser formula. The difference is that the Ganz-Griesser formula contains variable parameters, which are calibrated to a white standard, whereas the CIE formula is based on fixed factors.

Producers of white products decided that it would be most beneficial to deal with whiteness as a single-number index rather than using the three-colour-scale parameters. As a result, various whiteness indices were developed. In order for the whiteness index to be valid, it must be used on the type of materials for which it was intended. The index is most commonly used in textiles. When items are compared using a whiteness index, they must be similar in gloss, texture, translucency and other physical attributes.

4.11   Objective evaluation of fabric hand by KES and FAST systems

The two basic fabric requirements for the purpose of clothing are as follows:

i. Utility performance

ii. Comfort performance

The comfort performance of fabrics is of two categories:
i. Mechanical comfort

ii. Thermal comfort

The fabrics which have superior mechanical and thermal comfort performance are usually called high-quality fabrics. Thermal comfort is assessed from the permeability of the fabric to air, water and heat. Mechanical comfort is evaluated by a method called handle, which is traditionally done by visual and tactual impressions (it may vary from person to person, country to country). Mechanical comfort is evaluated by objective method; this expresses handle in terms of low-stress mechanical properties such as shear, tensile strength, compression and surface friction.

Mechanical comfort

Fabric handle: This is the level of individuals' response to touch when fabrics are held in hand. The fabric handle mainly depends on its low-stress mechanical properties. The low-stress mechanical properties of the fabric mainly depend on fibre, yarn and fabric structures and their properties.

Fabric handle or hand has traditionally been assessed by experts who arrive at an overall judgment on quality after manipulating the fabric with their hands. Professor Kawabata of Japan has carried out a great deal of work with the aim of replacing the subjective assessment of fabrics by experts with an objective machine-based system which gives consistent and reproducible results. Psychological responses of fabric handle are entirely determined by the physical and mechanical properties of fabrics. In particular, the properties of a fabric that affect its handle are dependent on fabric behaviour at low loads and extensions and not at the level of high load and extensions in which case fabric failure is likely.

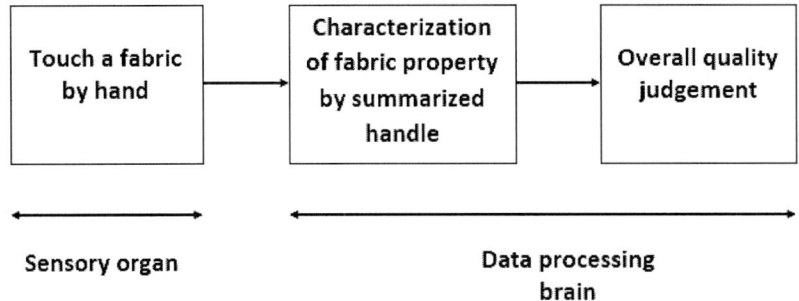

Figure 4.25 Subjective evaluation of fabric handle

Limitations of subjective evaluation of fabric

- End results are subjective and qualitative
- FSM restricts the scientific understanding of fabric hand

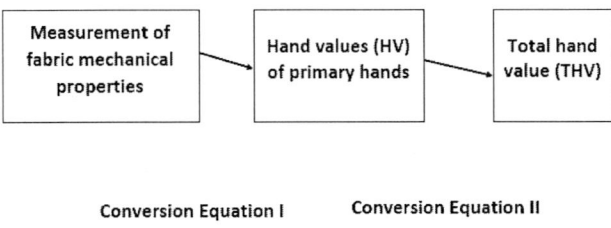

Conversion Equation I   Conversion Equation II

Figure 4.26 Objective evaluation of fabric hand

KES equipment

- KES-FB1: Tensile and shear properties
- KES-FB2: Bending
- KES-FB3: Compression
- KES-FB4: Surface friction and surface roughness

KES bending

- One edge of the sample is held by a fixed chuck, while the other is held by a moving chuck. A sensor-attached fixed jaw detects bending moment and curvature.

- The moving chuck follows a fixed orbit, turning its head at an angle so that a uniform curvature is maintained on the sample to find the relationship between the curvature and the bending moment.
- Clamp interval = 1 cm; rate of bending is 0.5 cm/sec.
- Maximum curvature is ±2.5 cm.

The following quantities are measured from the curve:

- Bending rigidity B = slope of the bending moment - curvature
- Moment of hysteresis 2 HB = hysteresis width of the curve.

KES compression

- Fabric sample is placed on the bottom plate of the instrument and a plunger is used to compress the sample at a constant rate.
- The fabric is compressed till a preset pressure is reached before starting the recovery cycle at the same constant rate.
- LC - linearity of compression thickness curve.
- WC - compressional energy (gf-cm/cm^2).
- RC - compressional resilience (%).

KES tensile

- The sample is clamped between the two jaws within the effective test area of 5 cm × 20 cm and is subjected to a constant tension of 10 gf-cm by a weight attached to the drum on which one of the jaws is mounted.
- During tensile testing, engaging the clutch arrests the free drum rotation and a torque detector connected to the drum is used to detect the tensile force in the fabric.
- Fabric is extended at a constant strain rate by moving the other jaw assembly until a pre-set load is reached before starting the recovery cycle.
- The tensile properties are measured by plotting the force extension curve between zero and a maximum force of 500 gf-cm (4.9 N/cm); the recovery curve that occurs as the sample is allowed to return to its original length is also plotted to give a pair of curves.
  - Tensile energy WT = the area under the load strain curve
  - Linearity LT = WT / area triangle OAB
  - Resilience RT = (area under load decreasing curve / WT) × 100%.

Shear test

In order to measure the shear properties a sample measuring 5 cm × 20 cm is sheared parallel to its long axis keeping a constant tension of l0 gf-cm (98.1 mN/cm) on the clamp.

The following quantities are then measured from the curve.

- Shear stiffness G ~ slope of shear force-shear strain curve
- Force hysteresis at shear angle of 0.5° 2 HG = hysteresis width of curve at 0.5°
- Force hysteresis at shear angle of 5° 2 HG5 = hysteresis width of curve at 5°.

Surface roughness

- Surface roughness is measured by pulling across the surface a steel wire 0.5 mm in diameter which is bent into a U shape.
- The contact force that the wire makes with the surface is 10 gf (98.1 mN).
- The value that is measured is SMD = mean deviation of surface roughness.

Surface friction

- Surface friction is measured in a similar way by using a contactor which consists of 10 pieces of the same wire.
- A contact force of 50 gf is used in this case, and the force required to pull the fabric past the contactor is measured.
- MIU = mean value of coefficient of friction.
- MMD = mean deviation of coefficient of friction.

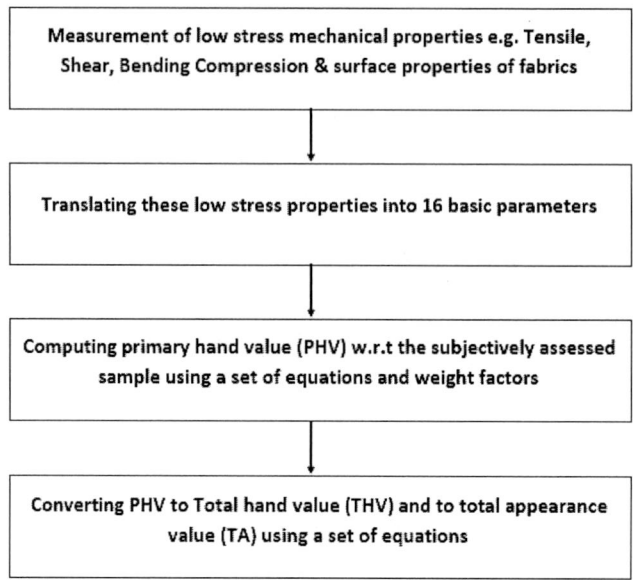

Figure 4.28 Basis of objective evaluation of KES-FB system

Table 4.3 Primary hand

KOSHI	Stiffness	A measure of crispness in bending
NUMERI	Smoothness	A measure of smooth, supple and soft fed
FUKURAMI	Fullness	A measure of bulk
SHARI	Crispness	A measure of crisp rigid fabric surface with a cool feel
HARI	Antidrape	A measure of flare

FAST system

- FAST - fabric assurance by simple test
- Designed by CSIRO for use by tailors and worsted finishers to highlight problems that may be encountered in converting a fabric into garment.
- Most of the parameters measured distinguish between loosely constructed fabrics which readily distort and the more tightly constructed ones which do not distort as easily.

The FAST system comprises four test methods:

- FAST 1 compression meter
- FAST 2 bending meter

- FAST 3 extension meter
- FAST 4 dimensional stability test

FAST compression

- The fabric thickness is measured on a 10 cm^2 area at two different pressures, first at 2 gf-cm^2 (19.6 mN/cm^2) and then at 100 gf-cm^2 (981 mN/cm^2) using the apparatus.
- This gives a measure of the thickness of the surface layer which is defined as the difference between these two values.
- The fabric is considered to consist of an incompressible core and a compressible surface.
- The fabric thickness measurements are repeated after steaming on an open Hoffman press for 30 seconds in order to determine the stability of the surface layer.

FAST – bending meter

- The bending rigidity, which is related to the perceived stiffness, is calculated from the bending length and mass/unit area.
- Fabrics with low bending rigidity may exhibit seam pucker and are prone to give problems in cutting-out.
- They are difficult to handle in an automated production line.
- A fabric with a higher bending rigidity may be more manageable during sewing, resulting in a flat seam but may cause problems during moulding.
- Bending rigidity - 9.8 x 10^{-6} MC^3 (μNm)

where C is bending length and M is mass per unit area.

FAST – extensibility

- The extension of fabric is measured in the warp and weft directions at three fixed forces of 5, 20 and 100 gf-cm (49, 196 and 981 mN/cm; size of the sample tested: 100 mm × 50 mm) using the apparatus.
- The extension is also measured on the bias in both directions but only at a force of 5 gf-cm (49 mN/cm): this is used for calculating the shear rigidity.
- Low values of extension give problems in moulding, produce seam pucker and give difficulties in producing overfed seams.

- High values of extension give problems in laying up, and such fabrics are easily stretched during cutting with a consequent shrinkage to a smaller size afterwards.

FAST – dimensional stability

- In order to measure dimensional stability, the fabric is dried in an oven at 105°C and measured in both the warp and weft directions to give the length L_1.

- It is then soaked in water and measured wet to give the wet relaxed length L_2. It is then dried again in the oven and measured again to give the length L_3.

- The following values for dimensional stability are then calculated from these measurements for both warp and weft:

Relaxation shrinkage = $[(L_1 - L_3) / LJ \times 100\%$

Hygral expansion = $[(L_2 - L_3) / L_3] \times 100\%$

- High values of shrinkage in a fabric produce problems in garment sizing due to shrinkage of panels; seam pucker may form in the final pressing stage.

- A small amount of shrinkage (usually below 1%) is required for fabrics intended to be pleated. A high value of hygral expansion can lead to loss of appearance in humid conditions as the fabric increases in dimensions under such conditions.

- The seams can also pucker in these conditions as the sewing thread prevents relative movement of the fabrics.

5

Testing for export market

Indian Textiles and clothing industry is the second large industry with a major contribution to the country's economy. The textiles industry accounts for 14% of industrial production, 4% of GDP; employs 45 million people and accounts for 12% of country's total exports. India is a major exporter of textiles and is less dependent on import. Most of the imports either raw materials or intermediate products that is made into the final product and exported. India ranks ninth among the major clothing exporters of the world, as per WTO, 2012. In case of textiles, India is the third largest exporter next to China and European Union.

Due to globalization the competition is getting tougher in the international market. In the export market, the three competitive factors that account are cost, quality and timely delivery. The quality, indeed, is checked by testing from the beginning (raw material) to end (final product). The buyers, who place order in India, want the manufacturer to test the materials at every stage and send the report for their approval. This testing has to be done at either the manufacturer's laboratory or second or third-party laboratory, an important condition being the testing laboratory is accredited by a suitable accreditation agency. Laboratory accreditation is discussed in Chapter 6.

5.1   Testing based on customer requirements

Based on the end-use, the buyer has a chart for testing the properties of fabric. For example, a basic T-shirt must be tested for the following parameters:

1. Weight (GSM)
2. Dimensional stability (Shrinkage)
3. Skewness (Spirality)
4. Bursting strength
5. Pilling test
6. Colour fastness to laundering, crocking, perspiration, water, light
7. Appearance after domestic laundering
8. Seam strength
9. Fibre analysis – the percentage of each fibre being reported in the label
10. Button attachment strength, if any

Fibre composition

Variety of textile fibres are available under various categories such as natural, regenerated and synthetic fibres. Any apparel made of textile fibres has to be labeled for fibre content since the nature, properties and price of each of these fibres are distinct. Most countries importing apparel and soft home furnishing products require fibre identification labels indicating the fibre type and percentage of fibre components. Some countries even use fibre composition to classify quota categories.

Qualitative and quantitative analysis of fibres employ the following methods:

- Microscopic appearance
- Flame test
- Staining test
- Infrared spectroscopy
- Thermal analysis
- Other methods

Table 5.1 Solubility test for fibre identification

FI-BRES	Ac-etone	60% H2SO4	70% H2SO4	5% NaOH	M-cresol	For-mic acid	DMF	Hcl	NaOcl
				REAGENTS					
Cotton	I S	I S	S	I S	I S	I S	I S	I S	I S
Silk	I S	S	S	S	I S	I S	I S	I S	S
Viscose	I S	S	S	IS	I S	I S	I S	I S	I S
Acetate	S	S	S	I S	I S	S	S	I S	I S
Nylon	I S	S	S	I S	S	S	S	S	I S
Acrylic	I S	I S	I S	I S	I S	I S	S	I S	I S
Wool	I S	I S	I S	S	I S	I S	I S	I S	I S
PET	I S	I S	I S	I S	S	I S	I S	I S	I S
Lycra	I S	S	S	I S	S	I S	S	I S	I S

S - SOLUBLE   I S - INSOLUBLE

Dimensional stability: It is the degree to which a material maintains its original dimensions when subjected to changes in temperature and humidity. Dimensional stability of textile material like fabric or garment is tested by measuring the shrinkage in fabric or garment after washing.

Shrinkage: Shrinkage of fabric may be defined as the contraction of fabric after washing and drying. All products made of textile fibres have a tendency to shrink and the shrinkage level is dependent on the fibre type.

Test specimen: Cut specimen, measuring at least 500 mm × 50 mm with edges parallel to the length and if the width of the fabric is less than 650 mm then full width specimen may be used and measurements made as per the requirement. If the fabric may possibly unravel during the test, overlock the edges of the specimen with dimensionally stable thread.

Marking: Place the specimen on the measuring table and make at least three pairs of marks on both length and width directions. Ensure that the distance between marks of each point is at least 350 mm, that no mark is placed within 50 mm from the edges of the specimen and that the measuring points are regularly spaced across the specimen.

Procedure

- Prepare the specimen and mark as described above and measure the dimensions before test (original length).
- Weigh the specimen before test and after washing. Place the material to be washed in the washer and add sufficient ballast and sufficient detergent to provide a good running.
- After the completion of hydro extraction of the washing procedure, remove the material, taking care that it is neither stretched nor distorted; weigh the material and immediately place the material and the ballast in the tumble dryer and turn on the dryer.
- Continue drying until the weight remains unchanged and then turn off the heat; allow running without heat for at least 5 minutes.
- Measure the dimensions of the material (final length) and calculate the shrinkage both length-wise and width-wise as follows:

Shrinkage % = [(original length – final length) / original length] × 100

5.2   Textile safety testing

Textiles and clothing is a common requirement of mankind and since the usage is very intimate to the skin the safety of textiles is most critical. The consumer demands for a high-quality product and the governmental regulations also insist the same. Textile testing is conducted to minimize risk and protect the interest of both manufacturers and consumers. At present, up to 7,000 chemicals are used in textile manufacturing, like colouring agents, fixing agents, special finishing agents and other chemical and reagents. Hence it is necessary to ensure the quality as well as safety of products by appropriate testing. Textiles impose many safety issues such as fire, fall accidents, choking, hazardous chemical content and so on. The safety of children's textiles and apparel is highly regulated by various agencies worldwide, which not only requires rigorous safety testing to ensure compliance to standards but also comprehensive performance testing to meet consumer expectations.

5.3   Chemicals – heavy metals, phthalates

The major chemical pollutants present on textiles are dyes containing carcinogenic amines, toxic heavy metals, pentachlorophenol, chlorine, halogen carriers, free formaldehyde, biocides, fire retardants and softeners. Heavy metals in textile wastewater is a major environmental problem and cause potential danger to human health when present in textiles.

Analysis of heavy metals

Both qualitative and quantitative analyses of heavy metals are carried out in textile industry. It is necessary to prepare the sample prior to analysis, as most samples are not ready for direct introduction into analytical instruments. In some cases, the analytes have to be released from the matrix by extraction or digestion in order to obtain a clear solution. The kind of sample preparation applied depends on the sample, the matrix and the concentration level of analytes; for instance, trace analysis requires more stringent sample preparation than major component analysis.

Commonly used extraction media are artificial sweat (pH 5.5 as well as pH 8.0) or saliva solutions by solid-phase extraction. For dissolving samples prior to metal analyses, the methods used are digestion in an open container; digestion in a pressurized, sealed container; microwave-assisted decomposition. Usually, samples are prepared by digesting the matrix using a strong acid.

Methods for the determination of heavy metals

After sampling followed by an appropriate sample preparation step, different analytical methods can be applied for qualitative and quantitative determination of heavy metals in, for example, wastewater samples, textile materials and reagent solutions.

Thin layer chromatography

Thin layer chromatography is a simple and quick procedure and it uses a screening method for determination of heavy metals present in textile materials. The greatest advantage of this method is its ability to separate and simultaneously identify a number of metal components present in or on the sample. As it demands only small amounts of samples and chemical reagents, it is highly favourable economically and ecologically. The disadvantage of this method is the relatively high limit of detection (LOD) for certain metals and its poor quantitative reproducibility.

Other methods being used are UV visible spectrometry, atomic absorption spectrometry (AAS), optical emission spectrometry with inductively coupled plasma (ICP-OES) and ICP mass spectrometer (MS).

Analysis of phthalates

Phthalates are a common contaminant. Even low levels of contamination can impact quantitative results. Avoid plastic materials and use only scrupulously cleaned glassware and equipment. All solvents should be tested for any phthalate content. Solvent blanks should be run through the GC-MS periodically to monitor for potential contamination. Disposable glassware is recommended where practical.

The procedure consists of three sections: sample preparation, extraction and analysis. In addition to the procedure described herein, certain alternate extraction and analysis methods are possible. Sample preparation method must be performed prior to the alternative extraction and analysis methods.

Sample preparation

Prior to analysis, each plasticized component part should be cut into small pieces (no dimension larger than 2 mm) or milled/ground into a representative powder. Each cut/milled plasticized component part will be considered a sample for testing described as follows. At minimum, prepare the amount required to constitute a sufficient sample size.

Phthalate extraction method

1. Testing lab shall determine how many replicate samples are necessary to meet quality assurance requirements. Prepare LRB concurrently with samples. Weigh out a minimum of 0.05 ± 0.005 g of sample into a sealable glass vial (weighed to an accuracy of ±0.5%); if sample is not uniform, collect more specimens to reduce sample variance.

2. Add 5 ml of THF to the sample. For samples weighing more than 0.05 g, add 10 ml of THF for every 0.1 g of sample (or a reasonable amount to dissolve sample*). Shake, stir or otherwise mix sample for at least 30 minutes to allow dissolution. * Sample may be sonicated and/or gently heated to expedite dissolution.

* Some materials may not dissolve completely. In this case, add an additional 2 hours to mixing time and then proceed.

3. Precipitate any PVC polymer with 10 ml of hexane for every 5 ml of THF used as described in Point 2 above. Shake and allow at least 5 minutes for polymer to settle (longer duration may be necessary to minimize clogging of filters in the subsequent step).

4. *Optional*: Filter THF-hexane solution through a 0.45 ^m PTFE filter. Collect a few ml of filtered solution in separate vial. This step is optional, but it is still recommended.

5. Combine 0.3 ml* of the THF-hexane solution with 0.2 ml of internal standard (BB, 150 ^g/ml) in a GC vial and dilute to 1.5 ml with cyclohexane.

* Depending on the phthalate concentration, a different dilution ratio may be necessary to produce results in the calibration range (e.g., if the sample is known to contain ~5% phthalate, increase the amount of filtered THF-hexane solution from 0.3 to 1.0 ml). For very low concentrations (~0.1%), use 1.3 ml of filtered solution. A pre-screen of the sample by IR spectroscopy can provide a broad estimate of phthalate concentration.

A GC-MS system with an auto-sampler is suggested for sample analysis.

5.4   Flammability – textiles, general wearing apparel and children's sleepwear

This test is used to measure and describe properties of natural or synthetic fabrics in response to heat and flame under controlled lab conditions. Any textile material can be evaluated using this test with the following exceptions: children's sleepwear, protective clothing, hats/gloves, footwear and interlining fabrics. Two factors are measured:

1. Ease of ignition (how fast the sample catches fire).

2. Flame spread time (the time it takes for the flame to spread a certain distance).

Sample preparation

1.1   Launderable/dry-cleanable fabrics: Cut 12 numbers of 6.5" x 2" samples (per group), Six from the original fabric (3 in warp direction and 3 in weft direction) and another six from the laundered/dry-cleaned fabric (3 in warp direction and 3 in weft direction)

1.2   Non-launderable/non-dry-cleanable fabrics: Cut 6 numbers of 6.5" × 2" samples (per group); 3 in warp direction and 3 in weft direction.

2. All fabrics are oven-dried for 30 minutes at 105°C.

3. All fabrics should then be placed in desiccator for at least 15 minutes before testing.

4. Secure samples into frame using two clips on each side.

5. Piled or napped fabric should be brushed with brushing device at this time, in order to raise the surface fibres.

Preparing the flammability tester

1. Main power switch is off.

2. Timer is set at zero.

3. Move rack to right, using the lever arm located on the front panel. Place sample holder in the instrument's sample racks such that the longest frame is on top.

4. Using the lever, slide the rack to the left until the sample comes in contact with the L-shaped locating arm. The burner tip now remains 5.16" away from the face of the specimen.

5. Fill the glass U-shaped manometer with water to adequate level (located on the left side of the chamber).

6. Stop cord of 50/3 mercerized cotton sewing thread is threaded through instrument.

Performing the test

1. Turn on the main power switch.

2. Select time of auto impingement (1, 5 and 10 seconds or manual). For the purpose of this class we will use the time of 5 seconds.

3. Press start button. Impingement is automatic and the flame is applied for a period of 5 seconds.

4. Timer will start and stop automatically, starting upon the application of the flame and ending when the weight is released by the burning of the stop cord.

5. Record the results and repeat the procedure for all the test specimens.

6. When testing is done, switch power off. Turn off gas supply.

Results and calculations

1. Calculate the arithmetic mean flame-spread time of the 6 (or 12) specimens.

2. If the mean time is less than 3.5 seconds, or any of the specimens do not burn, test 6 more specimens (or 12).

3. Calculate the arithmetic mean flame-spread time for all the 6 (or 12) specimens. Add all the 6 (or 12) specimens together and divide by 6 (or 12).

4. The time of flame spread is the average time for all the specimens of the sample material tested.

Interpretation of results

The following three classes are used to interpret results for a similar test:

Class I: These textiles are considered by the trade to be generally acceptable for apparel and are limited to the following:

1. Textiles that do not have a raised fibre surface but have an average time of flame spread in a test lasting 3.5 seconds or more.

2. Textiles having a raised fibre surface and have an average time of flame spread in a test lasting more than 7 seconds or burn with a surface flash (in less than 7 seconds) in which the base fabric is not affected by the flame.

3. Textiles for which no specimen ignites.

Class II: These textiles are considered by the trade to have flammability characteristics for apparel intermediate between Class I and Class III fabrics and are limited to the following:

1. Textiles have a raised fibre surface and have an average time of flame spread in the test lasting 4 to 7 seconds and the base fabric ignites, chars or melts.

Class III: These textiles are considered by the trade to be unsuitable for apparel and are limited to the following:

1. Textiles that do not have a raised fibre surface and have an average time of flame spread in a test that lasts less than 3.5 seconds.

2. Textiles having a raised fibre surface and have an average time of flame spread in a test lasting less than 4 seconds and the base fabric ignites, chars or melts.

Report

1. Report the standard test method used for testing.

2. Describe materials or products tested and method of sampling used.

3. Report average time in seconds of flame spread for fabric specimens that did ignite. If none of the specimens ignite, report DNI (did not ignite).

4. For raised surface fabrics, report the number of fabric specimens that ignites, chars or melts.

5. Report the number of specimens tested for each fabric.

6. Determine the classification of each fabric as defined above.

7. If fabric was laundered, report the information for the laundered samples and the non-laundered samples separately.

5.5   Labelling – fibre, fur and faux fur, care instructions, stuffed articles (law labels) and country of origin

The labelling of textile products harmonizes the names of textile fibres and other terms used in labelling or other documents accompanying these products, in order to ensure adequate information for consumers and to promote the sale of fabrics in the international market.

Clothing requires a trade description with the name of the country in which the goods were made or produced and a true description of the goods.

Textile products must be labelled or marked whenever they go into the market for production or commercial purposes. Where these products are not offered for sale to the end-consumer or when they are delivered in performance of an order placed by the state, labelling or marking may be replaced by commercial documents accompanying the export consignment. The names, descriptions and details of textile fibre content must be indicated in these commercial documents. They must also be indicated on products offered for sale to consumers.

With the exception of trademarks or the name of the undertaking, additional information required by this directive must be provided separately. EU member-states may require that their national language be used for the labelling and marking as required by relevant directives.

A textile product composed of two or more components which have different compositions must bear a label stating the fibre content of each component. Where two or more textile products have the same composition and form a single unit, they need to bear only one label.

Objectives of the textile labelling

- To protect consumers against misrepresentation in the labelling and advertising of textile fibre products
- To enable consumers to choose textiles on the basis of fibre content.

Labels containing fibre content; country of origin; identification of manufacturer, importer or dealer; and care instructions must be present at the time the end-user takes possession of the good. Labels containing care instructions must be attached permanently to the item.

Labelling requirements for products with fur

1. Animal name, according to the fur name guide
2. Name or registered identification number (RN) of the manufacturer, importer, marketer or distributor
3. Country of origin for imported fur written as 'Fur Origin: Country'
4. If the fur is mentioned, dyed, bleached or artificially coloured. If the fur product is composed in whole or substantial part (more than 10% of surface area) of pieces, such as paws, tails, bellies, sides, flanks, ears, throats, heads, scraps or waste fur
5. If the fur is used or damaged
6. The fibre content of any other materials used in the product.

Labelling requirements for products with faux fur

'Faux Fur' means artificially manufactured fur which is made to resemble real fur and not derived from animals.

All apparel items containing real or faux fur must have the terms 'real fur' or 'faux fur' listed on a label attached to the garment.

5.6   Mechanical hazards – drawstrings, small parts and sharp points and edges

Safety assessment of functional drawstrings

Purpose: To measure the length of functional drawstring used on children's garment.

Definitions

Functional drawstring – a cord, chain, ribbon, string or tape of any textile or non-textile material with or without embellishment, such as a toggle, pom-pom, feather or bead, which passes through a channel, loop(s) or eyelet(s) or similar openings, to adjust the size of the opening or part of the garment or to fasten the garment itself.

Functional drawstring measurement

Measure the functional drawstring from where the strings extend out of the garment when it is expanded to its fullest width. To ensure the garment is expanded to its fullest width, carry out the following steps:

A. For the non-stretchable garment and drawstring channel without elastic band and/or elasticity

i. The garment is held by lab technician in a direction perpendicular to the drawstring channel; the lab technician stretches the drawstring channel and ensures it is expanded to its full width.

ii. The garment is laid flat on the bench and it is ensured that there are no crinkles in the drawstring channel.

iii. The length is measured from the point where the drawstring exits the channel to the tip of string.

B. For stretchable garments and drawstring channel with elastic band and/or elasticity

i. The garment is laid flat on the bench. Measure and mark down the hip width.

ii. The first lab technician stretches the drawstring channel (waistband) to hip width (for woven) or hip width +1" (for knit); during the stretching, the exposed portion of the drawstring will be drawn into the drawstring channel by hand.

iii. While the garment is held under the tension at the specified width, a second lab technician measures the length of the drawstring's exposed portion with a steel ruler.

iv. The length is measured from the point where the drawstring exits the channel to the tip of string.

Report

Report findings, including type of drawstring, position of drawstring found on the garment, length of drawstring in inches and conclusion.

Sharp point test method

1. *Principle of operation*: The principle of operation of the sharp point tester is as follows. A rectangular opening 0.040 inch (1.02 mm) wide

and 0.045 inch (1.15 mm) long in the end of the slotted cap establishes two reference dimensions. Depth of penetration of the point being tested determines sharpness. If the point being tested can contact a sensing head that is recessed at a distance of 0.015 inch (0.38 mm) below the end cap and can move the sensing head a further 0.005 inch (0.12 mm) against a 0.5 pound (2.2 newton) force of a return spring, the point shall be identified as sharp. A sharp point tester of the general configuration or one yielding equivalent result will identify a sharp point.

2. *Procedure*

i. The sample to be tested should be held in such a manner that it does not move during the test.

ii. Part of the test sample may need to be removed to allow the sharp point testing device to test a point. Such dismantling of the test sample can affect the rigidity of the point in question. The sharp point test is performed with the point supported so that its stiffness increases to a level not greater than the stiffness of the point in the assembled sample.

iii. Using the general configuration, the adjustment and operation of sharp point tester can be carried out as follows: Hold the sharp point tester and loosen the lock ring by rotating it so that it moves towards the indicator lamp assembly at a sufficient distance to expose the calibration reference marks on the barrel. Rotate the gaging can clockwise until the indicator lamp lights. Rotate the cap counter-clockwise until an equivalent of five divisions (the distance between the short lines on the cap) has passed the calibration reference mark. Lock the gaging cap in this position by rotating the lock ring until it fits firmly against the cap. Insert the point into the gaging slot in all directions in which it is accessible and apply a force of 1.00 pound (4.45 newtons). A glowing light identifies the point as sharp.

iv. The test instruments used by the Commission in its tests for compliance with this regulation should have gaging slot opening with width and length no greater than 0.040 and 0.045 inches, respectively, and should have the sensing head recessed to a depth of no less than 0.015 inch. The force applied when inserting a point into the gaging slot should be no more than 1.00 pound.

Small parts testing

Small children tend to put almost everything in their mouth; thus it becomes necessary to avoid choking hazard in garments. This makes the testing of the pull strength of button and other small parts a major testing requirement in kids wear. Small parts and accessories, including bows, are thus put to a test with strength close to a force which a child cannot exert (normally to pull out a chalk button the force required is 90 newtons). There are several reasons a button or small part may fail the pull strength test; for example, a chalk button may fail pull strength due to insufficient number of stitches, button breakage. fabric rupture and so on. Thus, analysing the number of panels on which a small part is attached, ensuring even surface of attachment and use of right interlinings in lightweight fabrics become important after the test report is reviewed and deviation is spotted.

Accreditation of textile testing laboratory

6.1 Terminology

Accreditation: Accreditation is a formal, third-party recognition of competence to perform specific tasks. It provides a means to identify a proven, competent evaluator so that the selection of a laboratory, inspection or certification body is an informed choice.

Calibration: Calibration is a comparison between measurements – one of known magnitude or correctness and made or set with one device and another measurement made in as similar a way as possible with a second device.

Validation:

1. Validation is the confirmation by examination and the provision of objective evidence that a particular set of requirements for a specific intended use are fulfilled.

2. Assessing the degree to which an instrument accurately measures what it purports to measure, or a statistical technique or test that accurately predicts a value.

Reliability:

1. The ability of an apparatus, machine or system to consistently perform its intended or required function or mission, on demand and without degradation or failure.

2. Consistency and validity of test results determined through statistical methods after repeated trials.

6.2 ISO/IEC 17025

Any textile testing laboratory established for the purpose of testing needs to follow certain standard procedures, use proper testing equipments and also calibrate the measuring tools and equipments as per requirements. The test results reported by any testing laboratory must be reliable and valid. In order to provide general rules and procedures for a testing/calibration laboratory, an international standard was established in the year 1999 and revised in the year 2005. It is named as ISO/IEC 17025 General Requirements for the competence of testing and calibration laboratories. This international standard specifies the general requirements for the competence to carry out tests and/or calibrations, including sampling. It covers testing and calibration performed using standard methods, non-standard methods and laboratory-developed methods. The International Organization for Standardization (ISO) and the International Electrotechnical Commission (IEC) form the specialized system for worldwide standardization.

This international standard is applicable to all organizations performing tests and/or calibrations. These include, for example, first-, second- and third-party laboratories, and laboratories where testing and/or calibration forms part of inspection and product certification. This international standard is applicable to all laboratories regardless of the number of personnel or the extent of the scope of testing and/or calibration activities. When a laboratory does not undertake one or more of the activities covered by this international standard, such as sampling and the design/development of new methods, the requirements of those clauses do not apply. ISO/IEC 17025 covers several technical competence requirements that are not covered by ISO 9001:2000.

Implementing an ISO/IEC 17025 laboratory management system is a means to ensuring efficiency and technical competency in calibration and testing laboratories. A laboratory that establishes a laboratory management system compliant with ISO/IEC 17025 joins the growing world partnership of accredited laboratories. An ISO/IEC 17025 accreditation certificate will show potential customers that the laboratory values quality and that it has taken steps to ensure that testing results are accurate and reliable.

Quality has become an important concern to people all over the world. ISO/ IEC 17025 accreditation provides the assurance that calibration and testing laboratories are delivering good services and consistent data.

Benefits

- International recognition
- Sound management system
- Prevents defects
- Increased accuracy
- Cost savings
- Reduced waste
- Access to global marketplace

Steps to accreditation

1. File application
2. Receive quote

3. Enter contract

4. Prepare for accreditation

5. Document review

6. Preliminary assessment (optional)

7. Opening meeting (entry briefing)

8. Accreditation assessment

9. Closing meeting (exit briefing)

10. Corrective action

11. Executive committee's decision

12. Accreditation certificate

13. Surveillance assessments

A documented management system

- Defines the authority and responsibilities of personnel.

- Clearly communicates the objectives of the system and the laboratory's policies, procedures and work instructions.

- Promotes continuous improvement, which means the system is monitored regularly and changes can be incorporated easily.

- Ensures consistent performance.

The number of days required to complete an accreditation assessment depends on the following:

1. Size of laboratory

2. Number of employees

3. Complexity of calibration and/or testing operations

6.3   National Accreditation Board for Testing and Calibration Laboratories (NABL)

NABL is an autonomous body under the aegis of Department of Science & Technology, Government of India, and is registered under the Societies Act 1860. NABL has been established with the objective to provide government, industry associations and industry in general with a scheme for third-party assessment of the quality and technical competence of testing and calibration laboratories. Government of India has authorised NABL as the accreditation body for testing and calibration laboratories.

In order to achieve this objective, NABL provides laboratory accreditation services to laboratories that are performing tests/calibrations in accordance with ISO/IEC 17025:2005 and ISO 15189:2007 for medical laboratories. These services are offered in a non-discriminatory manner and are accessible to all testing and calibration laboratories in India and abroad, regardless of their ownership, legal status, size and degree of independence.

NABL accreditation system complies with ISO/IEC 17011:2004 and Asia Pacific Laboratory Accreditation Cooperation (APLAC) MR001. Based on evaluation of NABL operations by APLAC in 2000, NABL has been granted signatory member status by APLAC and International Laboratory Accreditation Cooperation (ILAC) under their Mutual Recognition Arrangements (MRAs). Under these MRAs, the reports issued by NABL-accredited laboratories are considered to be equivalent to reports issued by laboratories accredited by 76 accreditation bodies in 64 economies. NABL has undergone re-evaluation by a four-member APLAC evaluation team in July 2008. APLAC/ILAC has recommended extending NABL's mutual recognition arrangement (MRA) status for a further four years with recommendations for extending the scope of NABL testing capabilities to include medical testing laboratory in its purview as per new international standard ISO 15189:2007.

The users have access to information regarding accredited laboratories through web-based directory of NABL-accredited laboratories. NABL website is updated continuously with respect to status of accredited laboratories and their scope of accreditation. The list of laboratories which are either suspended or their scope of accreditation is partially or fully withdrawn is also available for the benefit of users. The laboratories will be able to acquire the necessary NABL documents through the website thereby eliminating postal delays. Suggestions are welcome from users of the website for further improvement.

Need for laboratory accreditation

The concept of laboratory accreditation was developed to provide a means for third-party certification of the competence of laboratories to perform specific type(s) of testing and calibration.

Laboratory accreditation provides formal recognition of competent laboratories, thus providing a ready means for customers to find reliable testing and calibration services in order to meet their demands.

Laboratory accreditation enhances customer confidence in accepting testing/calibration reports issued by accredited laboratories.

The globalization of Indian economy and the liberalization policies initiated by the government in reducing trade barriers and providing greater thrust to exports makes it imperative for accredited laboratories to develop competence that is at par with the competence of established international laboratories.

Benefits of accreditation

Formal recognition of competence of a laboratory by an accreditation body in accordance with international criteria has many advantages:

1. Potential increase in business due to enhanced customer confidence and satisfaction.

2. Savings in terms of time and money due to reduction or elimination of need for re-testing products.

3. Better control of laboratory operations and feedback to laboratories as to whether they have sound quality assurance systems and are technically competent.

4. Increased reliability on testing/calibration data and personnel performing work.

5. Customers can search and identify the laboratories accredited by NABL for their specific requirements from the Directory of Accredited Laboratories.

6. Users of accredited laboratories will enjoy greater access for their products, in both domestic and international markets, when their products are tested and certified by accredited laboratories.

Sources:

www.pjlabs.com
www.nabl-india.org

$$\frac{HBP}{RTK} \quad \frac{Shri \ Adhya}{20/09/25}$$

₹2,295

A-770